Skills
Techniques
and Decision Making

for Advanced
Level Geography

Skills Techniques and Decision Making

for Advanced Level Geography

Stephen Roulston   Mary Reid

Colourpoint Educational

6 5 4 3 2 1

© Roulston
Reid
2004

Designed by Colourpoint Books
Printed by The Universities Press
(Belfast) Ltd

ISBN 1 904242 12 X

Mary Reid, B.Sc (Hons), PGCE, is head of Geography in Dominican College, Belfast. She graduated from QUB and taught Geography in St Louise's Comprehensive College before her appointment to Dominican College in 1993. She has wide experience working with a number of examining boards but is currently a Principal Examiner for AS Geography with CCEA.

With gratitude and thanks to Tony, Catherine, Ellen and Thomas.

Stephen Roulston, B.Ed, MA, D.Phil, taught Geography briefly at Foyle and Londonderry College, and then moved to Ballymena Academy, becoming Head of the Geography Department there. In 1998 he took up a position as the Geography Field Officer with NEELB. He now works with C2k as a Learning NI Development Consultant.

To Irwin and Maureen, for all their help

Colourpoint Books
Colourpoint House
Jubilee Business Park
21 Jubilee Road
Newtownards
County Down
Northern Ireland
BT23 4YH

Tel: 028 9182 0505
Fax: 028 9182 1900
E-mail: info@colourpoint.co.uk
Web-site: www.colourpoint.co.uk

Contents

The A2 symbol indicates content relevant only to those studying the subject at A2 level

1 AS Fieldwork Report

Fieldwork is an essential component of the AS and A2 geography specification. If the study is well planned, focusing on a precise aim and conducted in an organised manner, with thorough follow-up, then geography fieldwork can become an exciting learning experience. It can enhance knowledge and understanding of the environment and help develop a more confident and competent approach to geographical studies.

In the AS3 examination a **fieldwork report** must be prepared and submitted for assessment along with a **table of data** collected in the field. The essential requirements for this table and report will be examined in detail but it is necessary for pupils and teachers to realise that they form an integral component of the fieldwork process. All stages of the geographical investigation **process** need to be followed through rigorously and reflectively in the classroom. This should allow maximum marks to be gained in Question 1 of the AS3 examination paper, which relates to all aspects of the fieldwork. The stages are outlined in Figure 1.

Figure 1 Schematic diagram illustrating the geographical investigation process

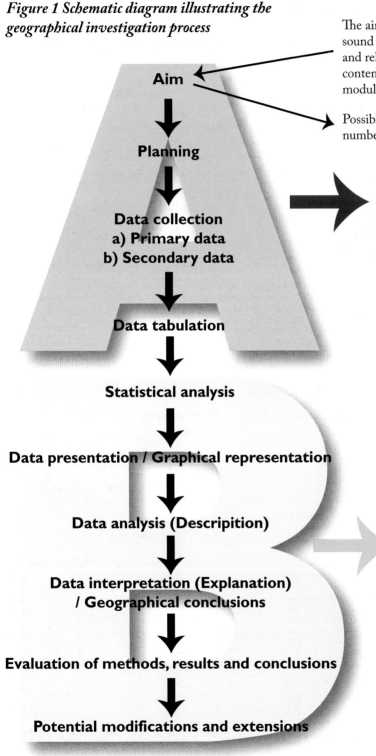

The aim must be based on sound geographical theory and relate to the specification content of AS1 or AS2 modular components.

Possibly sub-divided into a number of discrete hypotheses.

Only these elements are essential to the fieldwork report which is submitted for assessment.

Fieldwork follow-up which needs to be conducted in the classroom. These processes should not form any part of the fieldwork report submitted as part of the AS3 exam paper.

Aim

Planning

Data collection
a) Primary data
b) Secondary data

Data tabulation

Statistical analysis

Data presentation / Graphical representation

Data analysis (Description)

Data interpretation (Explanation) / Geographical conclusions

Evaluation of methods, results and conclusions

Potential modifications and extensions

1.1 The AS Fieldwork Report –
What should be included?

The aim

The aim may be a specific geographical question, problem or issue which should be stated clearly, concisely and should form the **focus** of the investigation.

It may be broken into a number of **sub-hypotheses** to investigate different aspects of the study and enable the geographer to manage and direct data collection.

Planning

Figure 2, on page 8, illustrates various aspects of pre-fieldwork planning which may be considered relevant to the chosen investigation.

Data collection

The data collection techniques employed in the field, or in the laboratory, should be described in detail in the report.

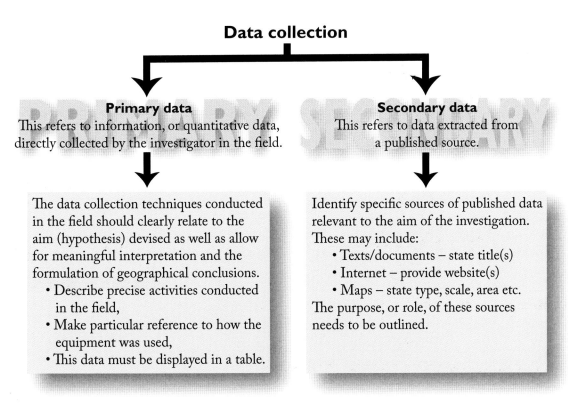

Data collection

Primary data
This refers to information, or quantitative data, directly collected by the investigator in the field.

Secondary data
This refers to data extracted from a published source.

The data collection techniques conducted in the field should clearly relate to the aim (hypothesis) devised as well as allow for meaningful interpretation and the formulation of geographical conclusions.
- Describe precise activities conducted in the field,
- Make particular reference to how the equipment was used,
- This data must be displayed in a table.

Identify specific sources of published data relevant to the aim of the investigation. These may include:
- Texts/documents – state title(s)
- Internet – provide website(s)
- Maps – state type, scale, area etc.
The purpose, or role, of these sources needs to be outlined.

Figure 2 : Fieldwork planning

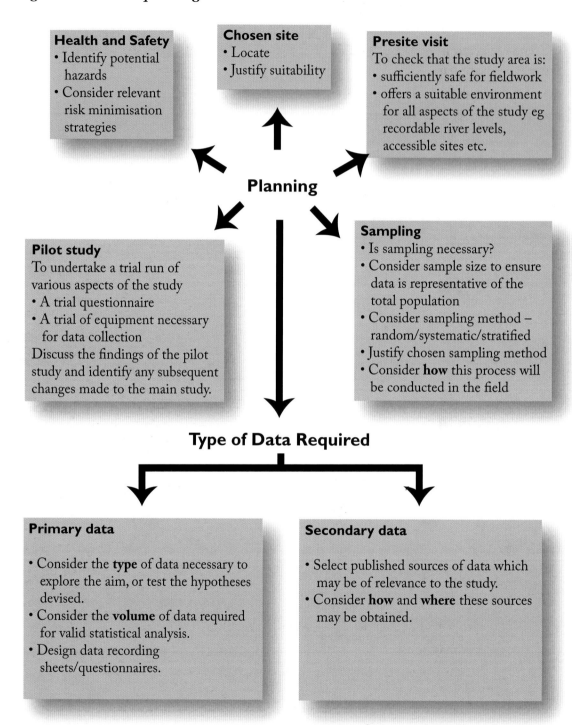

Health and Safety
- Identify potential hazards
- Consider relevant risk minimisation strategies

Chosen site
- Locate
- Justify suitability

Presite visit
To check that the study area is:
- sufficiently safe for fieldwork
- offers a suitable environment for all aspects of the study eg recordable river levels, accessible sites etc.

Planning

Pilot study
To undertake a trial run of various aspects of the study
- A trial questionnaire
- A trial of equipment necessary for data collection
Discuss the findings of the pilot study and identify any subsequent changes made to the main study.

Sampling
- Is sampling necessary?
- Consider sample size to ensure data is representative of the total population
- Consider sampling method – random/systematic/stratified
- Justify chosen sampling method
- Consider **how** this process will be conducted in the field

Type of Data Required

Primary data
- Consider the **type** of data necessary to explore the aim, or test the hypotheses devised.
- Consider the **volume** of data required for valid statistical analysis.
- Design data recording sheets/questionnaires.

Secondary data
- Select published sources of data which may be of relevance to the study.
- Consider **how** and **where** these sources may be obtained.

1.2 The Table of Data

The data collected in the field should be tabulated and submitted for assessment as part of the AS3 exam paper.

There are a number of essential requirements.

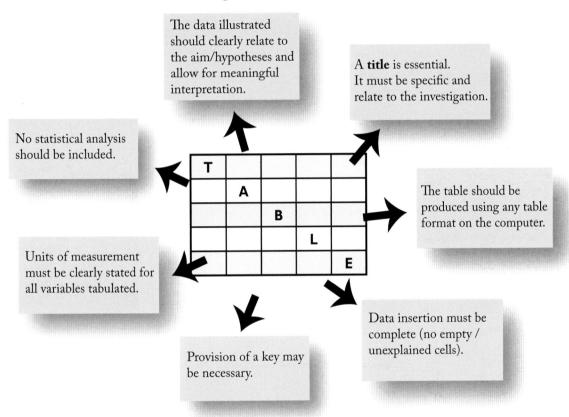

The data illustrated should clearly relate to the aim/hypotheses and allow for meaningful interpretation.

A **title** is essential. It must be specific and relate to the investigation.

No statistical analysis should be included.

The table should be produced using any table format on the computer.

Units of measurement must be clearly stated for all variables tabulated.

Data insertion must be complete (no empty / unexplained cells).

Provision of a key may be necessary.

Exam Tip

This table should be studied closely before the examination as its use will be required for parts of Question 1 on the AS3 exam paper.

You **may** be asked to:

- present the data using an appropriate **graphical technique** (refer to Section 4)
- analyse the data using an appropriate **statistical technique** (refer to Section 3)

Figure 3 Student Checklist

Report

	YES	NO
Have I structured my report adequately to include the Aim and detailed Planning and Data Collection sections?	☐	☐
Have I located and justified my study area?	☐	☐
Have I considered and fully justified any sampling method / methods which may have been employed?	☐	☐
Have I planned, edited, redrafted and proof-read my report to ensure that the quality of written communication is of a high standard?	☐	☐
Is my report within the 750 word limit?	☐	☐
Have I word processed my report using the specified font and text size (Times Roman 12 point)?	☐	☐
Is there a detailed discussion of all data collection methods for all variables illustrated in the table?	☐	☐
Have I excluded irrelevant material such as graphs, data analysis, interpretation, conclusions or evaluations?	☐	☐

Table of Data

	YES	NO
Have I provided a specific title?	☐	☐
Have I included units of measurement for all variables?	☐	☐
Have I produced my table using a computer?	☐	☐
Is my table complete and does it include all data discussed in the Data Collection section of the report?	☐	☐
Have I made sure that the table has no statistical analysis included?	☐	☐
Is there adequate data presented to allow for a thorough investigation of the stated aim and to allow for statistical testing?	☐	☐

The Fieldwork Report – What are the examiners looking for?

In order to gain top marks for this report, it may be advisable to focus on the marking criteria used by examiners to award a top level of attainment.

Assessment criteria – used by Examiners

A summary report which has planning and data collection sections **set out clearly** and which provides a **well structured** description of pre-fieldwork planning along with the process and methods of data collection. Quality of written communication is of a high standard. The report should not exceed the word limit (750 words).

Tips
- Plan sections relating to aim, data collection and planning
- Use sub-headings
- Devise paragraphs

Tips
The quality of written communication will be reduced by:
- spelling / grammatical errors
- poor sentence structure
- a lack of specialist geographical terms
- repetition between, and within, sections

Tips
To ensure that the word limit is adhered to, it may be necessary to edit, redraft and rewrite the report

Take Note

Once you have completed the checklist outlined in Figure 3 it would be advisable to seek the help and guidance of your teacher, or advisor, to gain a professional opinion on the quality of your work.

2 Data Collection

2.1 Sampling

Population is the term used to describe the complete set of items to be investigated. It is often impossible, impractical and unnecessary to study the entire 'population', so geographers commonly use the process of **sampling** to reduce the numerical quantity or areal coverage to be studied. Sampling is therefore employed to select a **portion** of the total population from which valid statistical inferences can be derived. This portion selected is commonly referred to as the **sample**. For example, if a geographer aims to investigate whether river variables change with distance downstream, it is impossible to study the entire course of the river. The geographer may decide to focus on a number of carefully selected sites which form the **sample**.

Sample Selection: key considerations

There are a number of important factors to be considered when deriving a sample population.

- **Sample Size** – The sample size must be sufficiently large to gain a **representative** view of the total population. Geographical conclusions will lack validity if they are based on an inadequate sample size. In statistical theory most data conforms to a normal distribution which, when plotted, produces a symmetrical bell-shaped curve (Figure 4). A large proportion of the values lie close to the mean with a decreasing proportion dispersed further from the mean. Geographers need to be aware that, for the sample population to be representative, it should reflect the characteristics of the normal distribution curve. This is more probable if the sample population is large. A high level of representation is achieved if the mean of the sample population closely approximates, or equates with, the mean of the total population.

Figure 4 The normal distribution curve

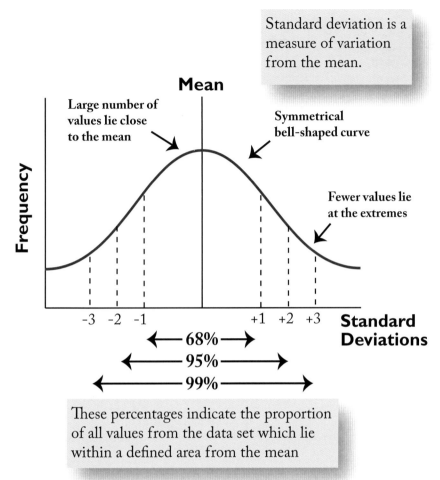

Standard deviation is a measure of variation from the mean.

Mean

Large number of values lie close to the mean

Symmetrical bell-shaped curve

Frequency

Fewer values lie at the extremes

-3 -2 -1 +1 +2 +3   **Standard Deviations**

←— 68% —→
←——— 95% ———→
←———— 99% ————→

These percentages indicate the proportion of all values from the data set which lie within a defined area from the mean

- **Elimination of bias** – To obtain an **accurate** and **representative** sample the researcher needs to eliminate bias by carefully considering the characteristics of the population. For example, a survey which investigates residents' perception of a proposed leisure complex in their locality would be biased and unrepresentative if the sample concentrated mainly on a specific age sector of the population.

- **Type of sampling procedure** – It is necessary to consider which type of sampling procedure would be most appropriate for the type of investigation. There are four main types (outlined in Figure 5), but geographers frequently use a combination of each to fulfil successfully the aim of their study.

Figure 5 Sampling procedures

Point sampling

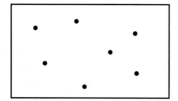

Individual **points**, or sites, are selected for investigation, eg sites along the course of a river, or a vegetation sampling quadrat which uses pins to locate points.

Grid or area sampling

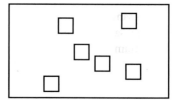

This involves delimiting an **area** for study. The size of the area varies according to the nature of the investigation, eg a quadrat is generally a $0.25m^2$ area used for vegetation sampling.

Line or transect sampling

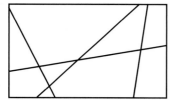

A **line** is plotted, along which data collection will occur. This is generally selected by geographers when variables under investigation are expected to demonstrate a progressive variation spatially, eg ecological succession studies, soil catena studies, etc.

Belt sampling

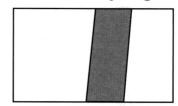

An **area** along a line is defined for investigation. This is also a useful method for vegetation and soil studies when the geographical aim involves a wider areal focus compared to transect studies.

- **Type of sampling method** – Researchers also need to consider the most appropriate sampling method to yield reliable information in relation to the aim of the enquiry. There are three types of sampling method – random, systematic and stratified sampling. These methods are summarised in Figure 6.

Figure 6 Types of sampling methods

	Random	Systematic	Stratified
Description	Random sampling involves selecting a sample using a random number generator, eg a random number table or the random number function on a calculator. Theoretically every member of the total population should have an equal chance of being selected. In spatial studies, sites can be located using the random numbers which can relate to co-ordinates which will inter-connect on a gridded map.	Systematic sampling involves the selection of data using a pre-determined interval. As with all sampling methods, this process can involve points, areas, lines or belts, eg it may involve interviewing every second householder along a street or studying soil at every 10m interval along a hill slope transect.	Stratified sampling is useful when sub-groups or subsets are clearly identified within the total population. This method should ensure proportional representation of each sub-group in relation to the total population, eg in a land use survey, where it is known that 70% of the study area is comprised of sandy soil and 30% clay soil, if 100 points are surveyed, 70 should be in the sandy zone and 30 on the clay.
Advantages	• The procedure is totally objective and should thus be unbiased. • If the sample size is sufficiently large, the data should be representative of the total population (reflecting the normal distribution). • Random numbers are easily generated for pre-fieldwork planning.	• This method is relatively simple, easy to employ and allows for well-organised data collection in the field. • It is particularly appropriate for studies which require an even coverage over time or distance, as it affords the researcher a degree of control over the data selection process.	• Conclusions are likely to be more valid when geographical sub-groups have been represented. • The method allows for flexibility as random, or systematic sampling can be used to select the data within the proportional sub-groups.
Disadvantages	• It does not take into account any underlying strata, or subsets, within the population. Therefore one grouping may be under, or over, represented. • The sampling can produce 'bunching' which is unsuitable for studies which aim to investigate a progressive spatial or temporal dimension.	• There is a higher chance of bias as the individual decides the interval and not all points/areas have an equal chance of selection. • It is possible that an underlying pattern could be missed and thus the total population may be misrepresented.	• If a multitude of sub-groups are inherent in the total population, then stratified sampling can become rather complex. • If strata are inappropriately identified, then the sample will not be truely representative and conclusions will be erroneous.

2.2 Vegetation sampling

Almost all ecological studies involve an element of vegetation sampling. Geographers widely use a **quadrat** to obtain a vegetation sample for measurement.

What is a quadrat?

A quadrat is a defined area of vegetation for ecological study. Quadrats vary in size depending on the type of vegetation unit. Although quadrats are traditionally, and commonly, square they can vary in shape.

Why use a quadrat?

There are many characteristics of vegetation which can be measured using the quadrat.

- **Vegetation cover** – this is the proportion of the ground occupied by a particular species, usually expressed as a percentage of the total quadrat area.

- **Vegetation frequency** – the chance of finding a species with one throw of a quadrat in a study area. If a species has a frequency of 50% it should occur in five of every ten quadrats analysed.

- **Biodiversity** – the number of species which occur within the defined quadrat area. This may be recorded as a numerical value or plant identification can provide more detail on the floral composition of the quadrat area.
  An alternative, but more subjective, assessment of vegetation abundance can make use of a devised key, which aims to simplify the data recording process. An example of a key devised by a group of students is illustrated in Figure 7.

Figure 7 Cover and distribution using combined code

	Cover		**Distribution**
+	Sparsely covered, barely visible	A	Isolated individuals
1	Cover evident, but less than 20%	B	Grouped or clumped
2	Plentiful, 21 – 50% of quadrat	C	Widespread patches
3	Numerous, 51 – 75% of area	D	Total carpet/single colonies
4	Cover is abundant, exceeding 75%		

COVER AND DISTRIBUTION
using a combined code eg 3B

Size of quadrat

Ecologists need to consider the size of the quadrat to ensure that the data is representative. The chosen quadrat size should be based on the size of the plant communities and their distribution pattern. Figure 8 illustrates recommended quadrat sizes for different plant communities.

Figure 8 Recommended quadrat sizes

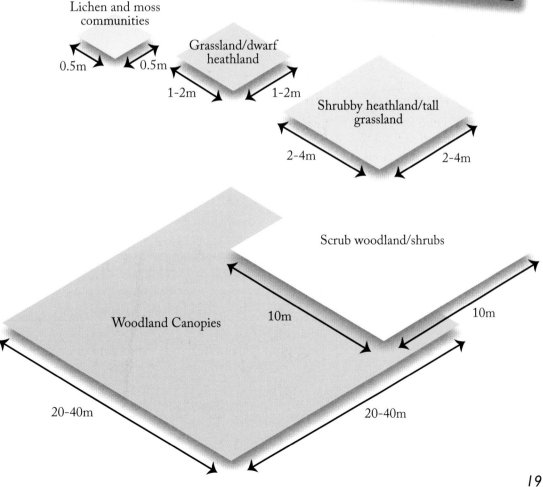

Lichen and moss communities

0.5m   0.5m

Grassland/dwarf heathland

1-2m   1-2m

Shrubby heathland/tall grassland

2-4m   2-4m

Scrub woodland/shrubs

10m   10m

Woodland Canopies

20-40m   20-40m

Types of quadrat

There are three main types of quadrat which are most frequently used in ecological studies. These are outlined in Figure 9.

Figure 9 Types of quadrat

	Frame Quadrat	Grid Quadrat	Point Quadrat
Characteristics	Frame quadrats are generally home-made, using four pegs carefully measured to represent the corners, with string to define the study area.	Grid quadrats are frames with the area sub-divided into 100 units to facilitate measurement.	A narrow frame with holes allows 10 pins through to indicate the points which record species as "hits". Unlike the grid and frame quadrat, it does not use an area approach. Pins can be adjusted to the height of the vegetation.
Advantages	• Useful when larger quadrat size is essential. • Do not flatten vegetation, producing a more accurate assessment of vegetation cover.	• The grid can make floristic assessment of cover more accurate. • Easily transported and quicker to use.	• Useful method to determine strata when vegetation is tall or multi-layered. • Vegetation is not flattened allowing for greater accuracy.
Disadvantages	• Estimations can be subjective without a grid to minimise error. • Difficult and time-consuming to construct.	• Tall vegetation can be crushed which can influence measurements.	• It does not allow for an assessment of cover, so the occurrence of minor species may be missed.

2.3 Questionnaires

What is a questionnaire?

- A questionnaire is a predetermined list of questions used by a researcher to obtain primary data relevant to the aim of a chosen study.
- It can be relatively simple, with questions to secure basic factual details, or it can be more complex, exploring public perceptions, attitudes and values relating to a geographical theme.

Designing a questionnaire

Questionnaire design involves considerable planning to ensure that questions are unambiguous and capture all aspects of information relevant to the chosen field study.

Take Note

The final analysis and conclusion is only as good as the data collected.

Plan and collect carefully.

Planning a questionnaire

Figure 10 Types of questions

Type of question	Advantages	Disadvantages
Closed These offer the respondent a fixed range of answers, usually with mutually exclusive options.	• Less time consuming to complete. • Answers lend themselves to quantitative analysis and the application of statistics.	• Unsuitable if public attitudes or perceptions are required. • They tend to be restrictive and inflexible as answers are predetermined.
Open Questions which offer the respondent the opportunity to outline their views or perceptions and formulate their own style of response.	• Respondents are is able to present their ideas without having to select from predetermined categories. Therefore data is likely to be more accurate.	• Difficult to collate and present in graphical or diagrammatic form. • Involves subjective interpretation by the researcher.
Likert Scales These are scales which allow the respondent to rank their attitude or view along a predetermined continuum.	• Numerical values can be produced, making graphical and statistical analysis possible. • Intensity/strength of feeling can be captured.	• Analysis can be complex and tedious if an expansive continuum is employed to cope with positive and negative views.

Question relevance

All questions included should capture information relevant to the aim of the study.

Length/Structure

The questions should be short, as respondents will be hesitant if the survey is time-consuming. The questionnaire should have well organised categories with clear, concise instructions.

Sampling

It may be necessary to consider:
1 Sample size – A larger sample size will ensure more reliable data.
2 Sampling method – It is important to choose the appropriate sampling method when administering the questionnaire eg random, systematic, stratified (refer to section 2.1).

What should be considered when designing and administering a questionnaire?

Avoid personal questions

It is not good practice to ask direct questions of a personal nature, eg income, age, religion etc.

Pilot study

A pilot study should be conducted on a small percentage of the sample population. This allows the questions to be 'road tested' for accuracy and ambiguity. Evaluation of the pilot study may make modification of the questionnaire necessary.

Avoiding bias

It is often necessary to consider factors which may influence the data. eg obtaining a cross-section of the population to avoid over-representation of a distinct group which would introduce bias.

Type of questions

The type of questions needs to be carefully chosen. There are possible alternatives which include – Open, Closed and Likert scales (see Figure 10).

Figure 11 Analysing questionnaire design

Visitor Survey

I am an A level geography student and I am conducting some research into human use of this coastal area. I would value your opinion and I would be grateful if you would take a few minutes to help me with my research.

Survey Number _____

Date _____

Time _____

Weather _____

Personal Details

Age: ☐ <20 ☐ 21-35 ☐ 36-50 ☐ 51-65 ☐ >65

Sex: ☐ Male ☐ Female

1. Have you ever visited this site before? ☐ Yes ☐ No

2. How far did you travel to get here today?
   ☐ <1km ☐ 5-9.9km ☐ >20km
   ☐ 1-4.9km ☐ 10-20km

3. How did you travel here today?
   ☐ Foot ☐ Car ☐ Train
   ☐ Bus ☐ Cycle ☐ Other (specify) _____

4. Did you visit the beach for any of the following reasons?
   ☐ Walking ☐ Dog walking ☐ Picnicing
   ☐ Horse riding ☐ Swimming ☐ Cycling
   ☐ Photography ☐ Sun bathing ☐ Sightseeing
   ☐ Other (specify) _____

5. Do you consider this area to be: (tick where appropriate)

	Excellent					Poor	
	5	4	3	2	1	0	
Beautiful							Ugly
Unspoilt							Spoilt
Quiet							Noisy
Clean							Dirty
Safe							Unsafe

6. Which management schemes from the list below have you seen today?
   ☐ Litter bins ☐ Education Boards ☐ Pathways
   ☐ Fenced off areas ☐ None ☐ Other _____

7. Would you like to suggest any improvements to the area?
   _____
   _____
   _____

Brief introduction to explain the purpose of the research.

Personal questions which can be completed by the researcher without asking the interviewee.

Begin with a simple, closed ended question.

Question 2 is an example of a **closed-ended question**. The respondent can select from the options provided.

Always include '**other**' to cover alternative possibilities.

Question 5 is an example of a '**Likert scale**' question. The respondent can tick the column to present the **strength** of their perception of the area.

Question 7 is an example of an **open-ended question**. The respondent can formulate their own answer.

3 Statistical Methods

Geographical data is often complex and lengthy, making interpretation difficult. It can be simplified and reduced into a concise mathematical form – this involves calculating **statistics**. Once statistics are clearly understood they are invaluable in helping to summarise, or interpret, aspects of geographical data.

3.1 Measures of central tendency

In any data set an average value can be obtained to represent the 'centre' of the distribution. The three most commonly used measures are:

• The **mean**

• The **median**

• The **mode**

The value of each statistic is outlined in Figure 13 , and its application demonstrated using the geographical data presented in Figure 12.

Figure 12

Geographical data

A group of students, carrying out a fieldwork river study, obtained the following data for bedload size, randomly sampled at one site along a river.

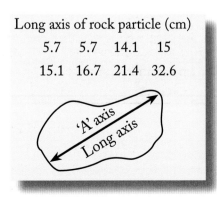

Long axis of rock particle (cm)

5.7	5.7	14.1	15
15.1	16.7	21.4	32.6

'A' axis
Long axis

Take Note

- All 3 measures of average have their own strengths and weaknesses which need to be understood if their 'usefulness' is to be fully appreciated.

- As they vary considerably within the same data set, the analysis and comparison of all three measures of average can provide a deeper understanding of the data.

Figure 13 Measures of Central Tendency – Summary Fact Sheet

Statistic	Application/ Calculation	Advantages	Limitations
Mean	The mean (or arithmetic mean) is obtained by totalling all the values making up the data set and dividing by the total number of values in the data set. Formula: $\bar{x} = \dfrac{\sum x}{n}$ where: $\bar{x}$ = mean $\sum$ = sum of x = value in data set n = number of values in data set	• It is an accurate measure of central tendency as all values are considered in the equation. • It is relatively simple to calculate and its value is commonly appreciated. • It is most reliable when there are many values in the data set and there are no abnormally extreme values.	• It is influenced by extreme values in the data set. • It may yield a decimal figure which can be inappropriate depending on the geographical context, eg an answer of 14.6 persons. • Reliability is reduced by small sample size.

Worked example

$\bar{x} = (\,5.7+5.7+14.1+15+15.1+16.7+21.4+32.6) \div 8$

$\bar{x} = 126.3 \div 8 = 15.7875$

Statistic	Application/ Calculation	Advantages	Limitations
Median	The median is the mid-point in a set of values when they are arranged in order of size (ranked numerically).	• It is a more preferable representation of the 'centre' of the distribution if the data contains values at the extremes. • It can be more reliable when there is 'bunching', or clustering of values in the data set.	• It is mathematically a less accurate average since the actual values do not form part of a calculation. • It is less reliable when there are few values in the data set or when there are large gaps between the values.

Worked example

5.7, 5.7, 14.1, 15, 〇15.1, 16.7, 21.4, 32.6

Median is 4½th value = 15.05   (15+15.1) ÷ 2

Mode	This is the value which occurs most frequently in the data set. A data set may have no identifiable mode.	• The mode does serve a useful purpose in describing the general shape of the distribution, eg unimodal (having one mode) or bimodal (having two modal classes) . • It is fast and easy to work out as there is no calculation involved.	• It is of limited value as a mathematical measure of central tendency as the modal value may represent extreme values within the data set, as indicated in the worked example. • It does not take into consideration the spread of the values within the data set.

Worked example – data is UNIMODAL

5.7, 5.7, 14.1, 15, 15.1, 16.7, 21.4, 32.6

Mode = 5.7

3.2 Measures of dispersion

The mean, as we have seen, summarises the 'centre' of a distribution in a relatively simple way. On its own, however, it is often not informative enough. **Dispersion** from the mean can provide a valuable insight into the data set and allows for fuller geographical comparisons.

Dispersion from the mean can be obtained by using three main statistical measures:

• The range

• The inter-quartile range

• Standard deviation

Although other measures of dispersion exist, only a knowledge of the **range** is essential.

Range

This is the simplest measure of spread – it is merely the difference between the lowest and highest values in the data set.

Advantages

• The range is easy to calculate and illustrates the **spread** of the data in the sample population.

• When used with the mean it can provide an insight into the distribution of values around the mean – it therefore develops the statistical usefulness of the mean.

Long axis of rock particle (cm)
5.7   5.7   14.1   15
15.1   16.7   21.4   32.6

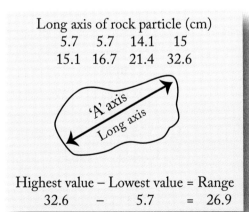

Highest value – Lowest value = Range
32.6   –   5.7   =   26.9

Limitations

• It is of limited use since it depends only on **two** values and ignores the distribution of all values in between. It is particularly disadvantageous if the extreme values are atypical.

• A general disadvantage is that the range of a sample population tends to increase as the sample size increases.

3.3 Correlation

Statistics, so far, have been concerned with single variable data sets, but in many investigations it is necessary to study the **relationship** between variables.

Key terms

Correlation – is used to describe the degree of association, or link, between two sets of variables.

Scattergraphs – can be used to graphically present the relationship between two variables. The types of relationship are illustrated in Figure 14.

Independent variable – or 'control' variable, is plotted on the horizontal 'x' axis of the scattergraph.

Dependent variable – the variable which is expected to be influenced by the control variable, is plotted on the vertical 'y' axis.

Line of best fit – this line can be plotted on a scattergraph to show the best possible trend. It can be estimated subjectively by eye but is much more accurate and objective when calculated mathematically.

Residual value – this is a point which lies at a distance from the line of best fit. It is often referred to as an **anomaly**.

Figure 14 Types of correlation

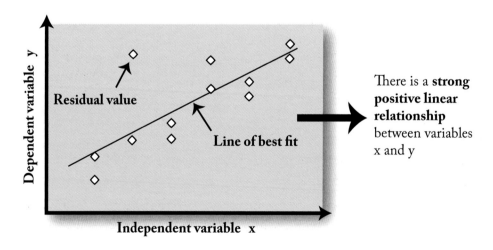

Residual value

Line of best fit

There is a **strong positive linear relationship** between variables x and y

Dependent variable y

Independent variable x

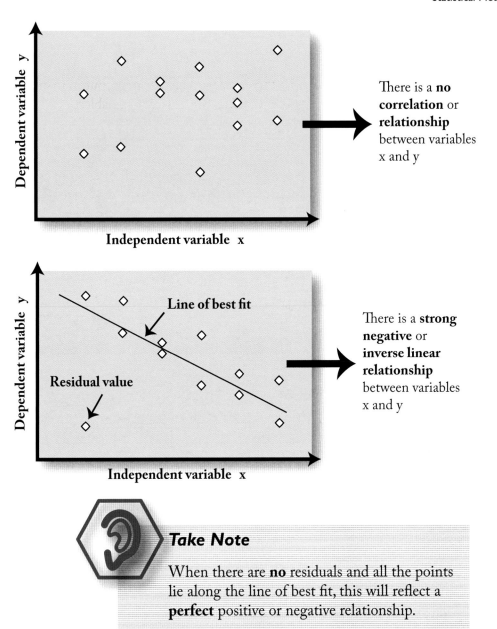

There is a **no correlation** or **relationship** between variables x and y

There is a **strong negative** or **inverse linear relationship** between variables x and y

Take Note

When there are **no** residuals and all the points lie along the line of best fit, this will reflect a **perfect** positive or negative relationship.

Spearman's rank correlation coefficient (r_s)

This is a statistical test which will provide a **reliable** measure of the **degree of association** between two sets of variables.

It will provide a numerical value which can be used to assess the **strength** of the linear relationship as well as the **type** of relationship, positive or negative, between the two sets of variables.

Calculation of r_s value

$$r_s = 1 - \left(\frac{6 \sum d^2}{n^3 - n} \right)$$

where
d = the difference in rank of the values of each matched pair

n = the number of ranked pairs

$\sum$ = the sum of

Worked Example

Hypothesis – There is a significant increase in river discharge downstream.

River site	x distance down-stream (km)	y River discharge (cumecs)	Rank x	Rank y	Difference in ranks d	d^2
1	0	0.26	12	11	1	1
2	0.4	0.15	11	12	-1	1
3	0.8	0.38	10	9.5	0.5	0.25
4	2.2	0.38	9	9.5	-0.5	0.25
5	2.6	1.62	8	8	0	0
6	3.0	1.93	7	5	2	4
7	3.4	1.95	6	4	2	4
8	3.8	1.92	5	6	-1	1
9	4.2	2.12	4	2	2	4
10	4.6	2.06	3	3	0	0
11	5.0	1.80	2	7	-5	25
12	5.4	2.70	1	1	0	0

$\sum d^2 = 40.5$

Steps

1. Rank the independent variable x from highest to lowest.
2. Rank the dependent variable y in a similar way.
3. Two or more identical values are given tied ranks.
4. Calculate the difference in ranks (d) for each pair of values.
5. Square each difference to remove all negative values (d^2).
6. Total all the d^2 values to obtain Σd^2.
7. Complete the calculation using the formula to determine the r_s value.
8. Interpret the rs value using the significance graph or table in Figure 15 (pages 31–32).

$$(r_s) = 1 - \left(\frac{6 \Sigma d^2}{n^3 - n} \right)$$

$$(r_s) = 1 - \frac{6 \times 40.5}{12^3 - 12}$$

$$(r_s) = 1 - \frac{243}{1716}$$

$$(r_s) = 1 - 0.14 = 0.86$$

Statistical interpretation

The **calculated** r_s value of 0.86 exceeds the **critical** value of 0.77 at the 99% significance level with 10 degrees of freedom. Therefore river discharge increases **significantly** downstream.

Interpretation of (r_s)

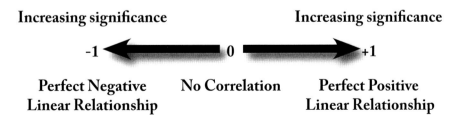

Increasing significance   Increasing significance

-1   0   +1

Perfect Negative   No Correlation   Perfect Positive
Linear Relationship   Linear Relationship

Therefore the r_s value will lie on a scale between +1 (perfect positive relationship) and −1 (perfect negative relationship). However we need to know whether any particular r_s value is close enough numerically to 1 to be considered significant.

Statistical significance

When investigating the link, or association, between two sets of variables, there is always the possibility that the relationship may have occurred by **chance**. Statisticians state that a greater than 5% possibility that the relationship occurred by chance is unacceptable to establish reliable significance. This means that the relationship may have occurred by chance more than 5 times out of 100. Therefore 5% is the level that geographers use as the **rejection level**.

How can the level of significance be determined?

In order to determine if the **calculated r_s value** is significant, and at what level, it is necessary to consult a **significance graph** or a **significance table**. (Figure 15)

Steps

1. Work out the **degrees of freedom** (n-2) and find this value on the horizontal axis of the significance graph.

2. Where the degrees of freedom line intersects with the 5% probability line on the significance graph – read across to the vertical axis to obtain the **critical rs value.**

3. If the **calculated** r_s value **exceeds** the **critical** r_s value, then the relationship is significant at the 5% probability level.

4. To check for a higher level of significance a similar method can be employed making use of the 1% and 0.1% probability lines to obtain critical values.

Figure 15 Spearman's rank correlation coefficient (r$_s$) Significance Graph

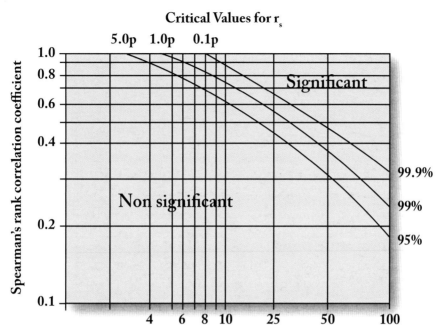

Degrees of freedom [Number of ranked pairs (*n*) −2]

95% significance level (5.0p)
This means that the relationship may have occurred by chance 5 times out of 100. This is a critical level of significance to achieve if the relationship is to be stated as significant with a acceptable degree of reliablity.

99% significance level (1.0p)
This means that the relationship may have occurred by chance only 1 time out of 100. This level, if achieved, provides a more reliable relationship than the 95% level.

99.9% significance level (0.1p)
This level indicates a highly significant relationship. There is only 1 chance out of 1000 that the association may have occurred by chance.

Spearman's rank correlation coefficient (r_s) Significance Table

Critical values of Spearman's rank correlation coefficient, r_s		
Significance Level		
Degrees of freedom	0.05 (5%)	0.01 (1%)
4	0.88	1.00
5	0.83	0.96
6	0.80	0.91
7	0.77	0.87
8	0.72	0.84
9	0.68	0.80
10	0.64	0.77
11	0.60	0.74
12	0.57	0.71
15	0.50	0.65
20	0.47	0.59
25	0.44	0.54
30	0.39	0.48
40	0.35	0.43
50	0.31	0.38

Take Note

- Spearman's Rank Correlation test can be used on data which consists of discrete variables (raw figures), or percentages, which can be ranked.

- The data should have at least 10 sets of paired values for the test to be statistically reliable.

- A significant value of r_s only indicates a strong degree of association between the variables. It does **not** prove that there is any **causal** link between them. In other words, the fact that two phenomena illustrate a positive, or inverse, trend is an indication that there **may** be a causative link which can be explored with further investigation.

3.4 Nearest neighbour analysis

The Nearest Neighbour statistic (Rn) is used to provide an objective measure of the distribution **pattern** of points arranged spatially in the landscape. In geography these points may relate to settlements, buildings, plants, springs, quarries, farms etc.

A numerical score is obtained from statistical analysis, which will allow the geographer to determine, with a degree of reliability, if the pattern is significantly
• **random**
• **clustered**
• **regular**

The calculated Rn numerical value will lie between **0** and **2.15**. Figure 16 provides an illustration of how the value can be interpreted.

We need to know if the values are close enough to **0, 1,** or **2.15** to indicate if the distribution pattern is significantly clustered, random or regular, with an acceptable degree of reliability. Therefore statistical analysis and interpretation are necessary to provide an objective and accurate conclusion.

Figure 16 Interpretation of Nearest Neighbour Value

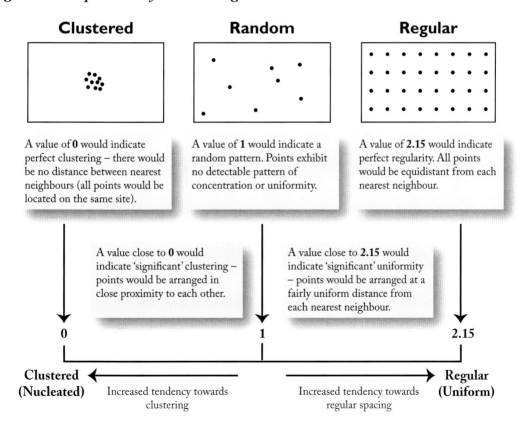

Clustered

Random

Regular

A value of **0** would indicate perfect clustering – there would be no distance between nearest neighbours (all points would be located on the same site).

A value of **1** would indicate a random pattern. Points exhibit no detectable pattern of concentration or uniformity.

A value of **2.15** would indicate perfect regularity. All points would be equidistant from each nearest neighbour.

A value close to **0** would indicate 'significant' clustering – points would be arranged in close proximity to each other.

A value close to **2.15** would indicate 'significant' uniformity – points would be arranged at a fairly uniform distance from each nearest neighbour.

0

1

2.15

Clustered (Nucleated) ← Increased tendency towards clustering

Increased tendency towards regular spacing → **Regular (Uniform)**

Worked Example

A geographer analysing the distribution pattern of post offices in Belfast (Figure 17 opposite), theoretically expected a random pattern. The expectation was based on the assumption that this low order service would not require a high threshold population to sustain economic viability and would therefore be located randomly in community, or neighbourhood, areas.

$$Rn = 2\bar{d} \sqrt{\frac{n}{A}}$$

where $\bar{d}$ = the mean distance between nearest neighbours

n = the number of points

A = the area under study

Steps

1. The points in the study area are accurately mapped and annotated with a number (Figure 17).

2. A table is constructed to record the straight line distance between each point (post offices in this example) and its nearest neighbour (Figure 18).

3. The average distance ($\bar{d}$) is calculated by totalling all the individual distance values and dividing by the total number of points.

4. The study area is then calculated (A).

5. The Rn numerical value is then calculated by substituting these values into the statistical formula (as illustrated in worked example Figure 18).

6. In order to interpret the Rn value, it is necessary to consult the Nearest Neighbour Significance Graph (Figure 19).

Figure 17 Post Office distribution in Belfast: Nearest Neighbour Analysis

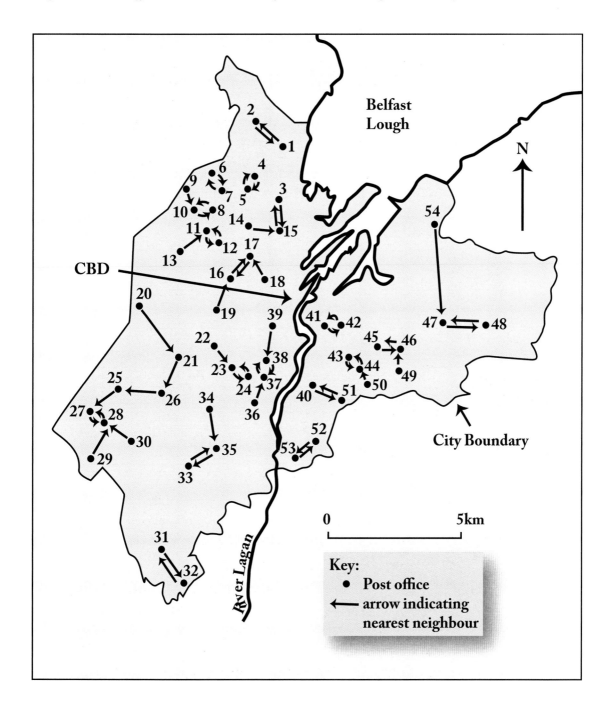

Figure 18 Post Offices in Belfast – Nearest Neighbours

Worked Example

Post Office	Nearest Neighbour	Distance (km)	Post Office	Nearest Neighbour	Distance (km)
1	2	1.22	28	27	0.44
2	1	1.22	29	28	1.22
3	15	1.00	30	28	1.00
4	5	0.44	31	32	1.33
5	4	0.44	32	31	1.33
6	7	0.67	33	35	1.11
7	6	0.67	34	35	1.22
8	10	0.56	35	33	1.11
9	10	0.67	36	37	0.78
10	8	0.56	37	38	0.44
11	12	0.44	38	37	0.44
12	11	0.44	39	38	1.11
13	11	1.11	40	51	1.00
14	15	1.22	41	42	0.33
15	3	1.00	42	41	0.33
16	17	0.89	43	44	0.33
17	16	0.89	44	43	0.33
18	17	1.00	45	46	0.67
19	16	1.00	46	45	0.67
20	21	2.22	47	48	1.44
21	26	1.22	48	47	1.44
22	23	0.89	49	46	0.67
23	24	0.44	50	44	0.44
24	23	0.44	51	40	1.00
25	27	1.11	52	53	0.67
26	25	1.44	53	52	0.67
27	28	0.44	54	47	3.56

Calculation of Rn numerical value

$$Rn = 2\bar{d}\sqrt{\frac{n}{A}}$$

$$Rn = 2 \times 0.902\sqrt{\frac{54}{144}}$$

$$Rn = 1.804 \sqrt{0.375}$$

$$Rn = 1.105$$

Area = 144 km²

Σd = 48.72

$\bar{d}$ = 0.902

Interpretation of Rn value

With 54 points a value of 1.105 indicates a **significantly random** distribution.

Figure 19 Nearest neighbour significance graph

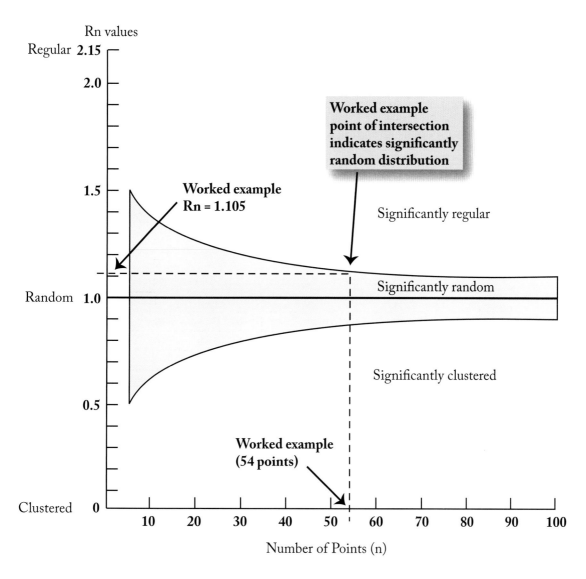

Locate the precise number of points on the x axis and the calculated Rn value on the y axis. Their point of intersection on the graph will determine whether the pattern can be classified as significantly:

• Random

• Regular

• Clustered

Take Note

Nearest Neighbour Analysis is a very useful statistic when analysing geographical distribution patterns, but the following points need to be considered.

- Statisticians state that a minimum of 30 points are required to ensure reliability and validity of conclusions.

- The delineation of the boundary of the area will require very careful consideration. A larger area will lower the Rn value and exaggerate the extent of clustering. Conversely the inclusion of a smaller area will increase the Rn value and exaggerate the extent of regularity. The total area is therefore critical to the statistical outcome.

- The measurement of the straight line distances between points and their nearest neighbours can create some problems. For example, if the points relate to settlements it may be difficult to locate the centre point for measurement, especially if the settlement has a linear morphology.

- The inability of the statistic to classify a linear pattern can be regarded as a limitation. For example, settlements are often distributed in a linear fashion as a result of relief, drainage, communications or other topographical landscape features. Therefore the application of nearest neighbour analysis may result in a misleading statistical interpretation.

- Although the nearest neighbour statistic provides a reliable classification of pattern (to a 5% probability level) it provides no explanation for the distribution. Geographers need to complete further research to explore potential geographical reasons for the patterns classified.

3.5 Chi-squared statistic A2

The Chi-squared test or χ^2 is used to find the significance of the difference between one frequency distribution and, usually, the frequency distribution expected by chance. It can only be used when the data are in the form of counted groups or frequencies. In other words, if you collected data and put them into classes (0-5, 5.1-10, 10.1-15 and so on) you would be able to compare these frequencies with those that would have occurred by chance.

Worked Example

For example, in a study of particle size along a river, if you had collected information from 3 points along a river, taking 30 particles from each site, measured the three axes and found the average and then grouped them into size categories, you might end up with a table like the one below.

In this case the fact that one of the particles at Point 1 had an average size of 1.7 cms is not recorded; it is classed as one of the six particles in this class at that point of the river. These are called **observed values**.

Particle size at three points on a river

Particle size/ River Points	Point 1	Point 2	Point 3	Total
>5cm	12	6	6	24
2.6 – 5cm	12	10	6	28
≤2.5cm	6	14	18	38
Total	30	30	30	90

At Point 3 there were six particles larger than 5 cms

Row total

Column total

Grand total

Hypothesis and null hypothesis

We want to know whether there is a difference between particle size at various points downstream. In order to ensure objectivity, a geographer will test a **null hypothesis**. This states that there are no differences (in our case in particle size at the different points sampled) other than might occur by chance. The symbol H_0 is used to abbreviate the null hypothesis in statistics. The null hypothesis here would be: '**There is no significant difference between particle sizes at various points downstream other than might occur by chance**'. We can also make a **hypothesis** (H_1) which in this case would be 'There is a significant difference between particle sizes at various points downstream'. It is always useful in Chi-squared tests to state the null hypothesis and the hypothesis before the test is carried out. It is always the null hypothesis which is tested – we are trying to prove ourselves wrong. This should produce a fairer result than if we were trying to prove ourselves correct.

Expected values

A Chi-squared test will compare this distribution of classes or frequencies, usually with what the distribution would have been if it had come about by chance. The chance figures for each cell of the frequency table are easily calculated. The table on page 39 shows that just 6 of the pebbles at Point 2 were over 5 centimetres in size. Since 24 out of the 90 particles in total were over that size, we would expect, if the results had occurred by chance, that there would be 8 pebbles of that size at points 1, 2 and 3.

To calculate the chance occurrence of each cell, multiply the row total by the column total and divide by the grand total eg (24 x 30) ÷ 90 = 8.0. This will provide the **expected values**.

Particle size at three points along a river with expected sizes according to chance

Particle size/River points	Point 1	Point 2	Point 3	Total
>5 cm	12 8.0	6 8.0	6 8.0	24
2.6–5 cm	12 9.3	10 9.3	6 9.3	28
≤2.5 cm	6 12.7	14 12.7	18 12.7	38
Total	30	30	30	90

Observed value

Expected value

$$\frac{38 \times 30}{90} = 12.7$$

Calculation

This is the Chi-squared formula.

$$\chi^2 = \sum \frac{(O-E)^2}{E}$$

The bigger the difference between the observed figures and the expected figures, the bigger the chance of the difference being found significant. For each cell calculate the expected figure and put it in the table in one corner of each cell. Apply the formula with the square of the difference between the observed and the expected figures divided by the expected for each cell and then add them together. The total figure provides the Chi-squared statistic, as the worked example below shows.

Chi-squared statistic $= [(12 - 8.0)^2 \div 8.0] + [(6 - 8.0)^2 \div 8.0] + [(6.0 - 8.0)^2 \div 8.0]$

$+ [(12 - 9.3)^2 \div 9.3] + [(10 - 9.3)^2 \div 9.3] + [(6 - 9.3)^2 \div 9.3]$

$+ [(6 - 12.7)^2 \div 12.7] + [(14 - 12.7)^2 \div 12.7] + [(18 - 12.7)^2 \div 12.7]$

$= 2.0 + 0.5 + 0.5$

$+ 0.78 + 0.05 + 1.17$

$+ 3.53 + 0.13 + 2.21$

$= 10.87$ (calculated value)

However there is one more step before this value can be used.

Degrees of freedom

Calculate the degrees of freedom for your table by multiplying the number of rows minus 1 by the number of columns minus 1.

$$df = (r-1) \times (c-1)$$

With three rows and three columns in the table on page 40, this would mean $df = (3-1) \times (3-1) = 4$. The higher the degrees of freedom value is, the higher the value of the result has to be to be significant.

Check the result

Now we can check whether the differences shown by our Chi-squared value are significant by using a significance table or graph such as the ones below.

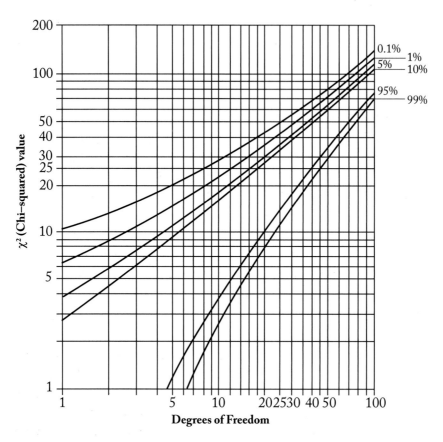

Critical values for Chi–squared test

Degrees of freedom	Significance level 0.05 (95%)	Significance level 0.01 (99%)
1	3.84	6.64
2	5.99	9.21
3	7.82	11.34
4	9.49	13.28
5	11.07	15.09
6	12.59	16.81
7	14.07	18.48
8	15.51	20.09

If the calculated value exceeds the critical value, the null hypothesis is rejected and the hypothesis is accepted: there *is* a significant difference between the data gathered and the chance values. Should it not exceed the critical value the null hypothesis is accepted: a significant difference does not exist.

Our calculated value of 10.87 exceeds the critical value of 9.49 at the 95% significance level (4 degrees of freedom), but not the 99% level, so we can say that our result is significant at the 95% level.

Significance levels

For the difference to be considered significant by geographers, the result must be higher than the critical value at the 0.05 or 95% significance level. What this means is that in less than 5 occasions out of 100 would the result arise because of chance. If the probability were less than 95 percent, it would not be considered high enough to be accepted as significant. If the 0.01 critical value is exceeded the result is even more significant, providing a 99% significance level. In that case such a result would occur in less than 1 out of every 100 tests where there is not a significant difference.

At the end of the test the null hypothesis can be revisited. We can reject ours as the result showed that there *was* a significant difference. In other words the null hypothesis – 'There is no significant difference between particle sizes at various points downstream other than might occur by chance' – can be rejected in favour of the alternative hypothesis **'There *is* a significant difference between particle sizes at various points downstream other than might occur by chance'.**

Take Note

Requirements for a valid Chi-squared text
- The data must be in counted occurrences for a number of categories.
- The total number of observations must be more than 20.
- No more than 20% of expected values should be less than 5.
- No expected value should be less than 1.

Exam Tip

It is possible that you could be required to complete a Chi-squared test in the A2 examination. However a much higher order skill would be the interpretation of a result and so you should be particularly prepared to do this. This requires knowledge of hypotheses and null hypotheses and of degrees of freedom and significance levels. You should also be able to describe real-life fieldwork where you have used Chi-squared to test for significant differences. This may be from physical geography, for example looking at changes in vegetation cover across sand dunes or from human geography, for example looking at changing land use with distance from the centre of the CBD.

3.6 Location Quotient

This statistical technique is used by geographers to measure the degree to which something is concentrated in one place compared to an area as a whole. It measures how much somewhere has more or less than its share of something. It is often used for studying the industrial geography of a region and also for ethnic composition of areas, but it has other uses too.

The formula is $LQ = \dfrac{RI}{RE} \div \dfrac{NI}{NE}$

where: RI = region's numbers of ethnic group or industry or whatever

RE = region's total numbers

NI = whole area's numbers of ethnic group, industrial employment or whatever

NE = whole area's total numbers

This sounds complicated until you work out a few. Let's try a simple one:

Worked Example

A town has 10,000 inhabitants and has 1,000 people working in textiles. One area of the town, with 1,000 inhabitants, has 100 people who work with textiles. The location quotient for this area of the town is 1.0 (work it out for yourself!). This means that there are exactly the number of people working in textiles in that area as you would expect, given the numbers working in textiles in the town as a whole. In other words, the small area has the same concentration as the whole town.

If the numbers working in textiles in the area we are studying were to rise to 200 (but the number working in textiles in the town were to remain at 1,000), the Location Quotient for our area increases to 2.0 – there are twice as many people working in textiles in that area than you would expect, given the concentration in the town as a whole. If we reduce the numbers on our study area to 50 the LQ drops to 0.5 showing a concentration of half that of the town as a whole.

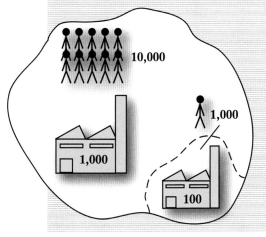

Worked Example

Below is a real example using 2001 Census data (www.nisra.gov.uk)

	Northern Ireland	Ballymena Local Government District	North Down Local Government District
Number of households	626,716	22,059	30,888
Number of households without central heating with sole use of bath/shower	28,892	1,295	1,121

LQ for Ballymena $= \dfrac{RI}{RE} \div \dfrac{NI}{NE}$

$= \dfrac{1,295}{22,059} \div \dfrac{28,892}{626,716}$

$= 0.0587 \div 0.0461$

$= 1.27$

LQ for North Down $= \dfrac{RI}{RE} \div \dfrac{NI}{NE}$

$= \dfrac{1,121}{30,888} \div \dfrac{28,892}{626,716}$

$= 0.0363 \div 0.0461$

$= 0.79$

This shows that Ballymena Local Government District has a higher concentration of houses without central heating with sole use of bath/shower than would be expected considering Northern Ireland as a whole. Similarly North Down Local Government District has a lower concentration than you would expect, given the concentration in Northern Ireland as whole.

These figures provide precise measures of concentration and allow geographers to start interpretation. What is it about North Down and Ballymena that produces the disparity?

Exam Tip

You should be prepared to calculate a Location Quotient in the A2-3 examination and you should be able to interpret a table of LQ results.

Preparing for A2-3 Section A

There is much you can do in preparing for this part of your A2-3 examination. You should:

- Revise ALL of your statistical techniques, including those from AS (Spearman's rank correlation, nearest neighbour, Chi-squared and Location Quotient).

- Remember the descriptive techniques such as measures of central tendency (mean, mode, median) and spread (range).

- Review the relevant parts of an up-to-date A-Level Geography specification which your teacher may have given to you, or which you can access from the CCEA site (www.ccea.org.uk).

- Do not omit data collection skills and techniques such as questionnaire design.

- Techniques of data presentation should also be covered.

- Reflect on fieldwork studies which you have done in your A-Level studies – there is often a question which asks you to use a fieldwork study in which you have been involved. This can be either a physical or a human geography fieldwork study.

4 Graphical Presentation and Mapping Techniques

4.1 Barcharts

A **barchart** or bargraph is one of the simplest and most commonly used methods of geographical representation. Data is displayed in **columns** or **bars** the height of which represents the **quantity** of a discrete component. Each component is measured on the vertical axis using a scale and therefore this method of presentation is useful when comparing quantities eg rainfall values throughout a year.

How to construct a barchart

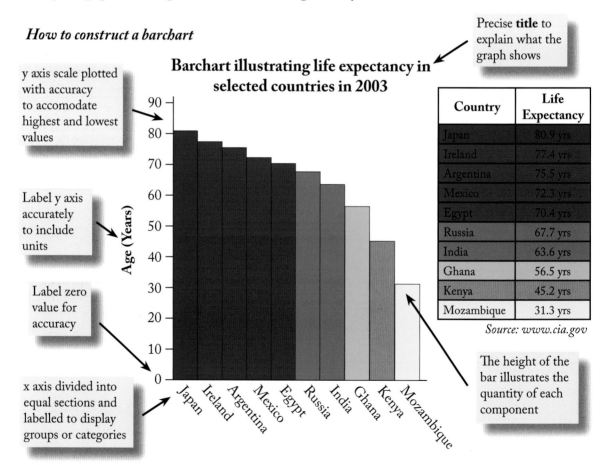

y axis scale plotted with accuracy to accomodate highest and lowest values

Label y axis accurately to include units

Label zero value for accuracy

x axis divided into equal sections and labelled to display groups or categories

Precise **title** to explain what the graph shows

Barchart illustrating life expectancy in selected countries in 2003

Age (Years)

Country	Life Expectancy
Japan	80.9 yrs
Ireland	77.4 yrs
Argentina	75.5 yrs
Mexico	72.3 yrs
Egypt	70.4 yrs
Russia	67.7 yrs
India	63.6 yrs
Ghana	56.5 yrs
Kenya	45.2 yrs
Mozambique	31.3 yrs

Source: www.cia.gov

The height of the bar illustrates the quantity of each component

Barchart variations

There are a multitude of variations of the bar chart commonly used to display geographical data. Three examples are shown.

A Reverse Bars

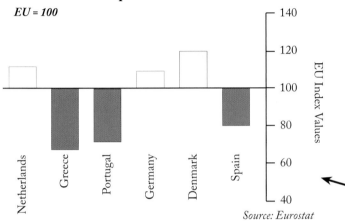

GDP per capita in selected European countries compared with EU base

EU = 100

EU Index Values

Source: Eurostat

Country	GDP per capita EU index value
Spain	80
Denmark	120
Germany	108
Portugal	73
Greece	66
Netherlands	113

This design is an interesting variation which provides a visual comparison of GDP in relation to the EU base line. In this example bars above the base line (EU = 100) illustrate countries which exceed the EU index value. Bars below the base line have GDP values less than the EU index.

B Population Pyramids

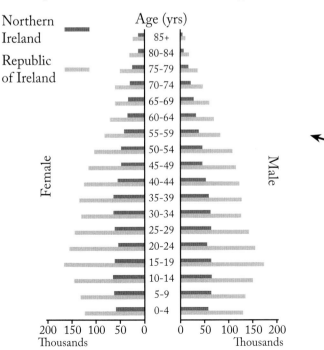

Population Structure - N Ireland & Rep Ireland

Northern Ireland

Republic of Ireland

Age (yrs)

85+
80-84
75-79
70-74
65-69
60-64
55-59
50-54
45-49
40-44
35-39
30-34
25-29
20-24
15-19
10-14
5-9
0-4

Female

Male

200 150 100 50 0
Thousands

0 50 100 150 200
Thousands

Population pyramids, commonly used in demography, are specially designed barcharts which allow geographers to compare age and gender variations in the population.

Source: Statistical & Research Agency and Central Statistics Office

51

C Compound or Component Barcharts

These 'divided' barcharts display the **proportional contribution** of various elements within the total quantity. Components can be displayed as numerical values or as percentages and their relative differences can be visually compared. These bars can be drawn horizontally or vertically.

How to construct a compound barchart

Employment patterns for selected LEDCs and MEDCs				
	Country	Primary Industry (%)	Secondary Industry (%)	Tertiary Industry (%)
LEDCs	Nepal	92	1	7
	Bangladesh	57	10	33
	Mali	85	2	13
MEDCs	UK	3	24	73
	USA	3	21	76
	Germany	4	30	66

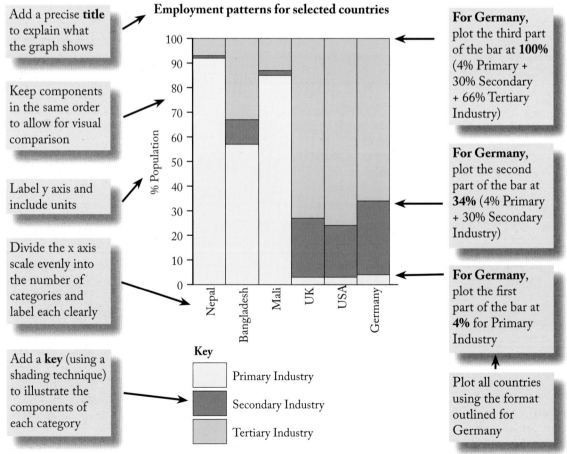

Add a precise **title** to explain what the graph shows

Keep components in the same order to allow for visual comparison

Label y axis and include units

Divide the x axis scale evenly into the number of categories and label each clearly

Add a **key** (using a shading technique) to illustrate the components of each category

Employment patterns for selected countries

% Population

Key
Primary Industry
Secondary Industry
Tertiary Industry

For Germany, plot the third part of the bar at **100%** (4% Primary + 30% Secondary + 66% Tertiary Industry)

For Germany, plot the second part of the bar at **34%** (4% Primary + 30% Secondary Industry)

For Germany, plot the first part of the bar at **4%** for Primary Industry

Plot all countries using the format outlined for Germany

4.2 Line Graphs

Line graphs are appropriate when data is continuous. They show how the **dependent** variable (plotted on the y / vertical axis) changes in relation to the **independent** variable (plotted on the x / horizontal axis). They display trends, or changes, which can be absolute or relative. A simple line graph shows how one variable changes, while multiple line graphs show the changes of many variables plotted using the same axes. If the data shows a large range, eg 5 – 50 000, then log or semi-log graph paper will be more suitable.

How to construct a line graph

Population Change in Northern Ireland and Republic of Ireland (1841 – 2001)

Year	Northern Ireland population (thousands)	Republic of Ireland population (thousands)
1841	1649	6528.8
1861	1396.4	4402.1
1881	1304.8	3870
1901	1237	3221.8
1911	1250.5	3139.7
1926	1256.6	2972
1946	1313.8	2955.1
1961	1425	2818.3
1981	1543	3443.4
2001	1689.3	3838.9

Add a specific **title** to explain what the graph shows

Plot and label the **dependent** variable on the y axis. Include units of measurement

Join up points or crosses with a ruler to produce a line

Plot accurately a dot or a cross where the two variables intersect

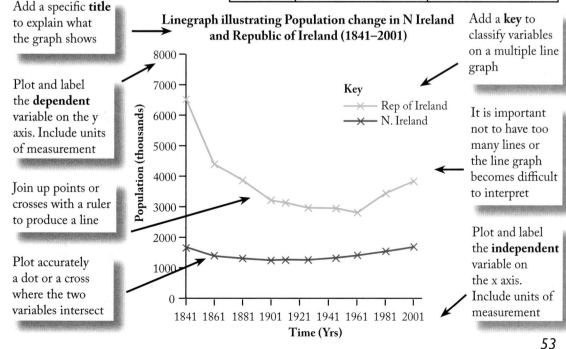

Linegraph illustrating Population change in N Ireland and Republic of Ireland (1841–2001)

Key
✕ Rep of Ireland
✕ N. Ireland

Add a **key** to classify variables on a multiple line graph

It is important not to have too many lines or the line graph becomes difficult to interpret

Plot and label the **independent** variable on the x axis. Include units of measurement

Take Note

The data on the x axis is **continuous** ie progression over time, and therefore a line graph is an appropriate graphical presentation technique.

4.3 Piecharts

A **piechart** or piegraph is simply a circle divided into segments which are proportional to the component categories of the data set.

Piecharts are appropriate when the total can be divided into component parts. The circle of the piechart represents 100% and therefore the segments or components should collectively total 100%.

Piecharts can be drawn using raw data or percentages calculated for each component. To construct a piechart it is necessary to calculate the size of each sector in **degrees**. The example shown below illustrates the calculations and method necessary.

Guyana: Ethnic Groups

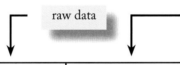
raw data

To calculate the number of **degrees** divide the population of each ethnic group by the total population and multiply by **360°**.

A circle is **360°** and therefore the number of degrees is proportional to each segment.

Ethnic Group	Number of people (millions)	Size of piechart segment (degrees)
East Indian	29.8	183.6
African	16.94	104.4
Mixed	8.19	50.4
Amerindian	2.4	14.4
European/Chinese	1.17	7.2
Total	58.5	360°

Example: The number of degrees for the European/Chinese sector

$$= \frac{1.17}{58.5} \times \frac{360}{1} = 7.2°$$

Take Note

If values are recorded as percentages, then multiply by **3.6** to obtain the number of degrees for each sector. A circle is 360° and each percent is one hundredth of the circle.

Therefore 1% is equal to:-

$$360 \div 100 = 3.6^O$$

How to construct a Piechart

Title to explain what piechart shows

Radial line used to begin plotting the first sector

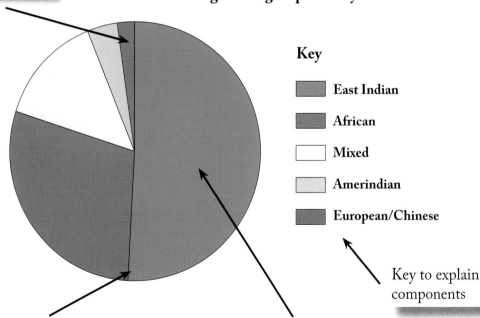

Piechart illustrating ethnic groups in Guyana

Key

- East Indian
- African
- Mixed
- Amerindian
- European/Chinese

Key to explain components

Plot the first sector by measuring 183.6^O using a protractor and drawing the radius

It is advisable to start with the largest category and then work clockwise around the circle in descending order of size

55

4.4 Proportional graphs A2

Proportional graphs use the area of the symbols to represent the sizes of a variable eg unemployment rates. The most common shapes used include circles, squares and triangles. A square which is twice the height and twice the width of another square will represent **four** times the value of the other. Often circles are used in proportional graphs with the area of the circle proportional to what is being shown. For example, Scottish visitors to Northern Ireland spent 12 times more than Welsh visitors in 2001 (£48 million compared to £4 million). If these figures were being drawn on a proportional circle then the **area** of the Scottish circle should be 12 times greater than the Welsh one.

Sometimes these circles are placed on a map to provided *located* proportional circles. Often these circles are divided, so that for example visitors of different nationalities are represented within the proportional pie charts showing different visitor numbers. While this can be an effective graphical technique, it is often difficult for the viewer to distinguish the difference between the numbers represented in a small segment of a large 'pie' and those in a large segment of a small 'pie'.

Source of tourists in Northern Ireland 1972, 1986 and 2001

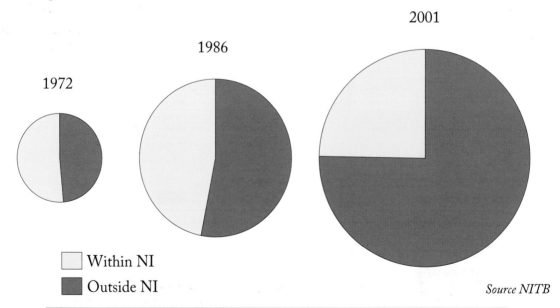

Source NITB

Year	From outside NI	From within NI	Total
1972	435,000	457,000	892.000
1986	824,100	727,600	1,551,700
2001	1,676,000	548,000	2,224,000

Care should be taken when interpreting proportional graphs as they can also be extended into 3-D symbols such as spheres or cubes. In this case the volume of the symbol is what represents the values shown on the graph. Thus a cube which is twice the width, height and length represents eight times the value of the other cube. Occasionally the graph may use a symbol such as a bag of grain to represent the data. The viewer has to be clear whether the data are being drawn on a volume or an area basis. Sometimes a close examination of these graphs, even from reputable sources, betrays a lack of understanding by those who have produced the graph – a good reason for using a trained cartographer to draw maps and graphs!

Numbers of visitors to selected NI tourist attractions (2001)

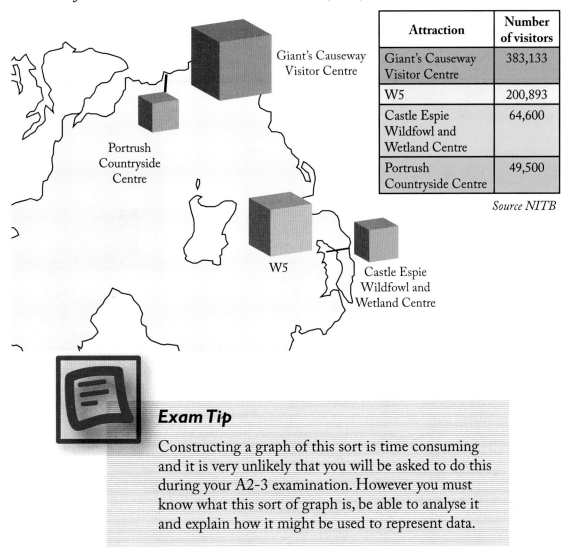

Attraction	Number of visitors
Giant's Causeway Visitor Centre	383,133
W5	200,893
Castle Espie Wildfowl and Wetland Centre	64,600
Portrush Countryside Centre	49,500

Source NITB

Exam Tip

Constructing a graph of this sort is time consuming and it is very unlikely that you will be asked to do this during your A2-3 examination. However you must know what this sort of graph is, be able to analyse it and explain how it might be used to represent data.

4.5 *Mapping Techniques:* Choropleth maps

A choropleth map is one which is shaded to show the relative **density** of an area. The gradation of colour reflects the range of values and the deepest most intense colour illustrates the highest values. The shading indicates visually contrasting distribution patterns between regions. Although relatively easy to interpret, constructing a choropleth map can be much more challenging.

How to construct a choropleth map

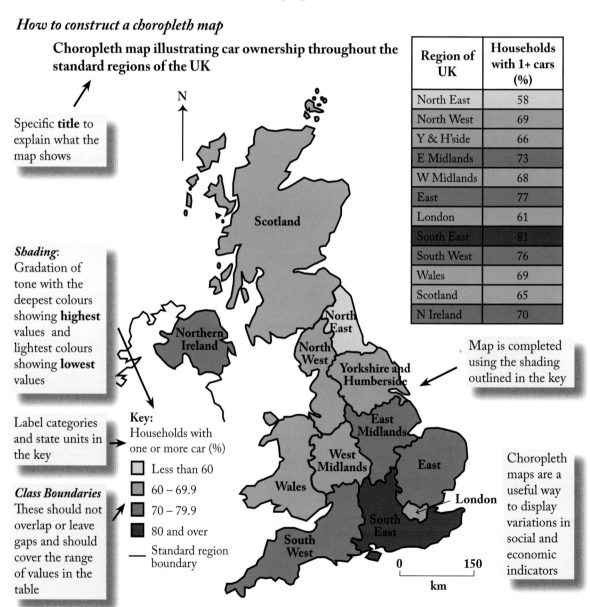

Choropleth map illustrating car ownership throughout the standard regions of the UK

Region of UK	Households with 1+ cars (%)
North East	58
North West	69
Y & H'side	66
E Midlands	73
W Midlands	68
East	77
London	61
South East	81
South West	76
Wales	69
Scotland	65
N Ireland	70

Specific **title** to explain what the map shows

Shading: Gradation of tone with the deepest colours showing **highest** values and lightest colours showing **lowest** values

Label categories and state units in the key

Class Boundaries These should not overlap or leave gaps and should cover the range of values in the table

Key:
Households with one or more car (%)
- ☐ Less than 60
- ☐ 60 – 69.9
- ☐ 70 – 79.9
- ☐ 80 and over
- — Standard region boundary

Map is completed using the shading outlined in the key

Choropleth maps are a useful way to display variations in social and economic indicators

N

Scotland

Northern Ireland

North East

North West

Yorkshire and Humberside

East Midlands

West Midlands

East

Wales

London

South East

South West

0   150
km

To produce a choropleth map, cartographers (map drawers) need to:

- **Devise suitable classes.** The data must be grouped into classes or categories. It is important to consider the number of classes necessary to cover the entire range of values present. Four to six groups is often quoted as desirable. Suitable class boundaries need to be established but class intervals do not need to be uniform. It is important that boundary values do not overlap and it is equally important to ensure that gaps do not occur between classes.

- **Consider appropriate shading.** Shading should be graded from dark to light. Darkest tones indicate highest values while lightest tones show lowest values. This can be very effective visually when varying tones of one colour are used.

An alternative method of shading

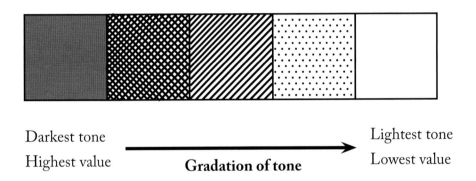

Darkest tone — Gradation of tone → Lightest tone

Highest value — Lowest value

Take Note
Limitations of choropleth mapping

Although choropleth maps provide geographers with a striking visual image of contrasting distribution patterns, it is important to recognise that:

- Oversimplification will have occurred. They present an area or region as having a uniform value range and therefore fail to present possible intra-regional variations.

- Maps often provide striking contrasts at regional boundary zones, which can be unrealistic.

4.6 *Mapping Techniques:* Dot distribution maps

A dot map is an effective and relatively simple mapping technique used to display a geographical **distribution**.

Each dot represents a specific value and their distribution and density on a map allow a geographer to interpret patterns.

Dots can be distributed randomly within defined 'regions' or alternatively they can be plotted in actual locations if the information is available.

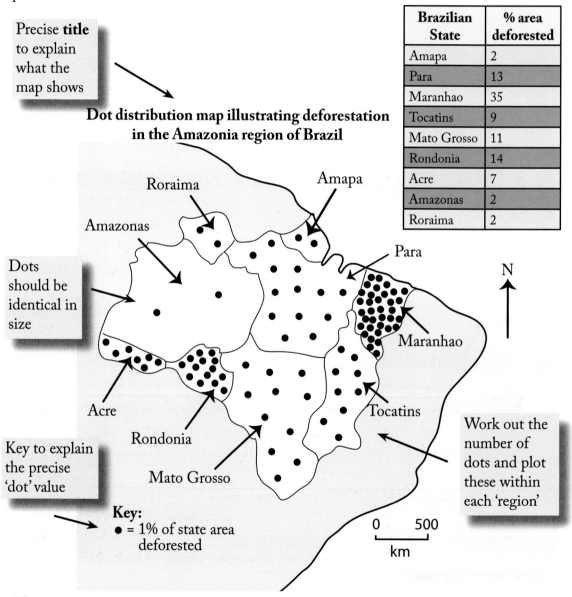

Precise **title** to explain what the map shows

Dot distribution map illustrating deforestation in the Amazonia region of Brazil

Brazilian State	% area deforested
Amapa	2
Para	13
Maranhao	35
Tocatins	9
Mato Grosso	11
Rondonia	14
Acre	7
Amazonas	2
Roraima	2

Dots should be identical in size

Key to explain the precise 'dot' value

Work out the number of dots and plot these within each 'region'

Key:
● = 1% of state area deforested

0   500
km

4.7 *Mapping Techniques:* **Flow line maps**

Flow lines are drawn to display visually the **volume** and **direction** of movement between two defined locations.

The **width of the line** is proportional to the quantity of movement, and the **arrow** indicates the direction of flow. They can be used effectively to represent a wide range of geographical information including traffic, migration and tourist flows.

How to construct a flow line chart

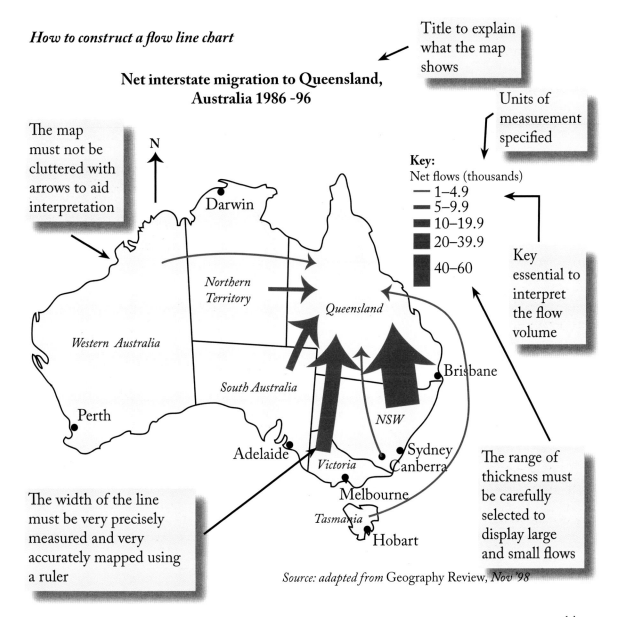

Title to explain what the map shows

Net interstate migration to Queensland, Australia 1986 -96

Units of measurement specified

The map must not be cluttered with arrows to aid interpretation

Key:
Net flows (thousands)
— 1–4.9
— 5–9.9
— 10–19.9
— 20–39.9
— 40–60

Key essential to interpret the flow volume

The width of the line must be very precisely measured and very accurately mapped using a ruler

The range of thickness must be carefully selected to display large and small flows

Source: adapted from Geography Review, *Nov '98*

4.8 *Mapping Techniques:* **Annotated sketch maps**

The sketch maps here used the 1:50,000 map accompanying the White Strand golf course Decision Making Exercise (Module 5: 2001).

A sketch map emphasising physical features

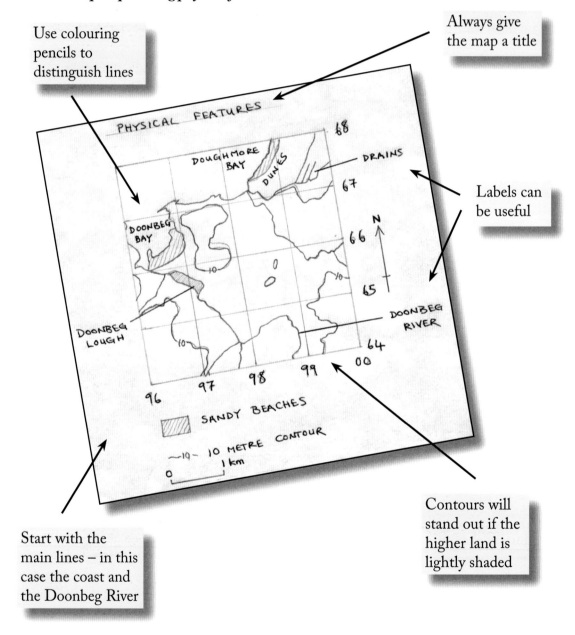

Use colouring pencils to distinguish lines

Always give the map a title

Labels can be useful

Start with the main lines – in this case the coast and the Doonbeg River

Contours will stand out if the higher land is lightly shaded

A sketch map emphasising human features

Only add the detail that you are required to

Draw the grid lines lightly to act as guides

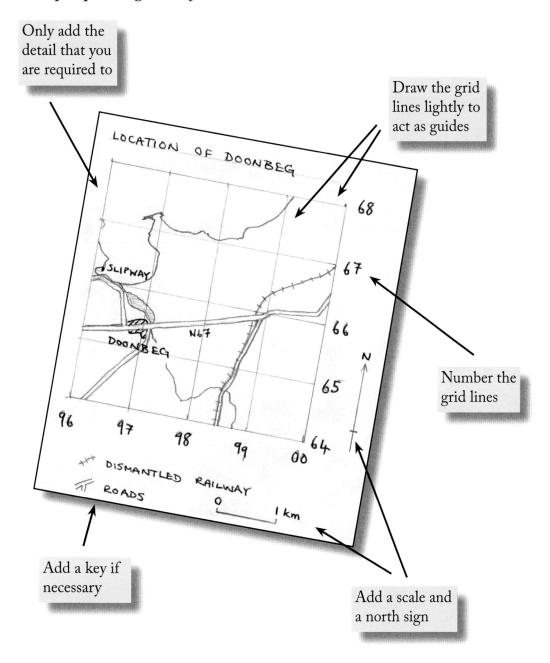

Number the grid lines

Add a key if necessary

Add a scale and a north sign

4.9 *Mapping Techniques:* **Isoline maps**

Isolines are lines which join together points of the same value. Examples are:

- isotherms, joining places of the same temperature together
- isobars, joining places of the same air pressure together
- isohyets, joining places with the same rainfall together
- contours, joining places with the same height above sea level together

Diagram A

The interpolation between points

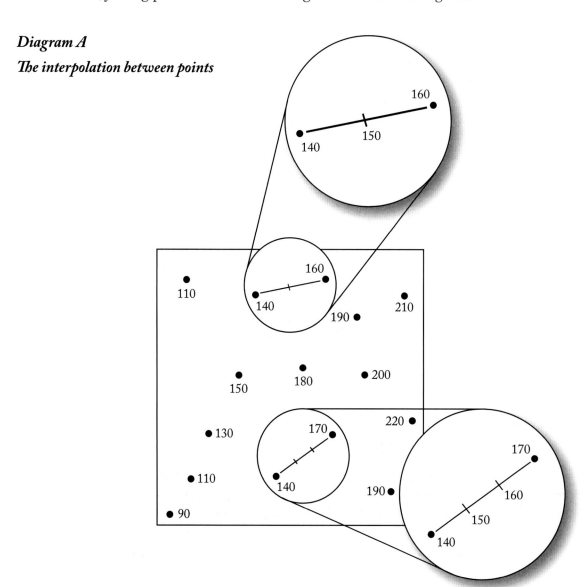

Mapping isolines involves interpolation (meaning *between two points*). A meteorologist may know the annual rainfall in Point A is 1150 mm and at Point B is 850 mm. He can then interpolate the value of 1000 mm exactly half way between them. When a number of such values are known the points can be joined to give isohyets (see diagrams A and B).

In interpolation the assumption is made that the change between two points is steady, and it may not be. Nevertheless we can be sure that there **must** be a 1000 millibar value between a point with 980 Mb and another with 1010 Mb but it may not always lie exactly two thirds of the way between these points. However any inaccuracies so produced will generally be small and the more points used to create an isoline map the more accurate the result will be.

Diagram B
The interpolation
joined up

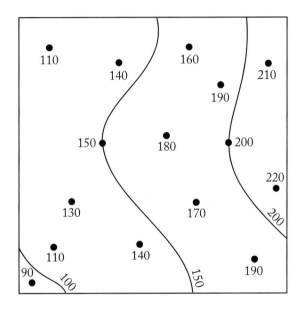

Exam Tip

Constructing isoline maps by interpolation is time consuming, so it is most unlikely that you will be required to do one in A2-3. Nevertheless you may have to complete one. You must be familiar with the technique and its purpose and be able to interpret an isoline map provided for you in the examination.

5 Decision Making

The Decision Making Exercise (DME) in A-level Geography has changed little since it was first introduced in 1988. You will be presented with a variety of resources related to a real-world contemporary geographical issue. Recent issues included the proposed building of a waste incinerator in County Meath (2004), a new container port in Southampton (2003), a fifth terminal for Heathrow Airport (2002) and a golf course in County Clare (2001). In each case the candidates needed to use their geographical skills to examine conflicting viewpoints about the issue. They were required to take on a rôle, usually that of a planner or government minister, and using their geographical skills had to set the issue in context and describe the implications of each position. Finally a decision was made in relation to the proposed development and was justified.

The types of resources which will be presented in the Resource Booklet are unlikely to change much from year to year. There will be a question, usually the third question in the booklet, a resource booklet and usually a separate map.

5.1 Written resources

There will be a number of written resources. One may give a background to the issue and there may be information from the main parties with an interest in the proposed development. A context for the issue is provided in the question paper and this often contains information which should be used. Generally there are a number of quotations summarising the views of some of the parties involved.

Maps

There are usually a range of maps showing the location of the development and more detailed maps showing the development in more detail. There is often a separate Ordnance Survey map, generally 1:50,000 but sometimes 1:25,000. These maps always have a key attached so there is no requirement to learn all the symbols associated with these maps.

Photographs

These are also often included in Decision Making Exercises and can be useful to help you to form a picture of the area. Occasionally there are photographs of models of the proposed development or artist's impressions showing what the development may look like if it were to go ahead. It is often possible to refer to these photographs, such as the one below, in your answer. This photograph shows a sign reflecting the beliefs of those opposed to a waste incinerator in County Meath and comments in your answer could relate to the sign, to the road traffic or to the landscape of the area.

Graphs or diagrams

These are sometimes used to illustrate complex data related to the issue and are likely to be important sources of information for your answer. The graph below, for example, appeared on the 2002 paper about a proposed fifth terminal for Heathrow Airport. It shows the loudness of individual noise events at Hounslow, near to the airport. The graph could have been used to back up the case of those opposing another terminal as loud noises begin to be commonplace at 4.30 am. Equally, loud noises could be argued to be relatively infrequent between 11.00 pm and 6.00 am and this could have been used in evidence by those in favour of the new terminal. This graph was an important resource in the arguments in relation to Terminal 5.

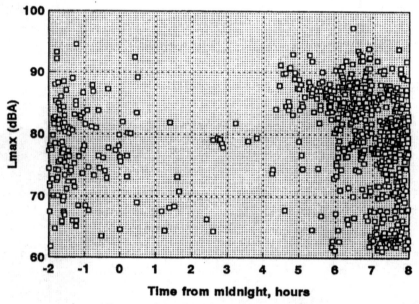

Source: Dept of Transport (1993) Report of the Field Study of Aircraft Noise and Sleep Disturbance

Tables

There are often one or more tables of data which you may be required to redraw as a graph and incorporate into the report.

Not all issues will require all of these resources, but most will be present each year. Do not neglect any of the resources. All resources included in the DME should allow you to support or to counter some point of view or to provide a context for the issue. This is particularly true of photographs and other illustrations which you should refer to if you can.

This should be an easy two marks – make your headings and sub-headings clear

Adopt the role and maintain it but do not overdo it – it is worth two marks only

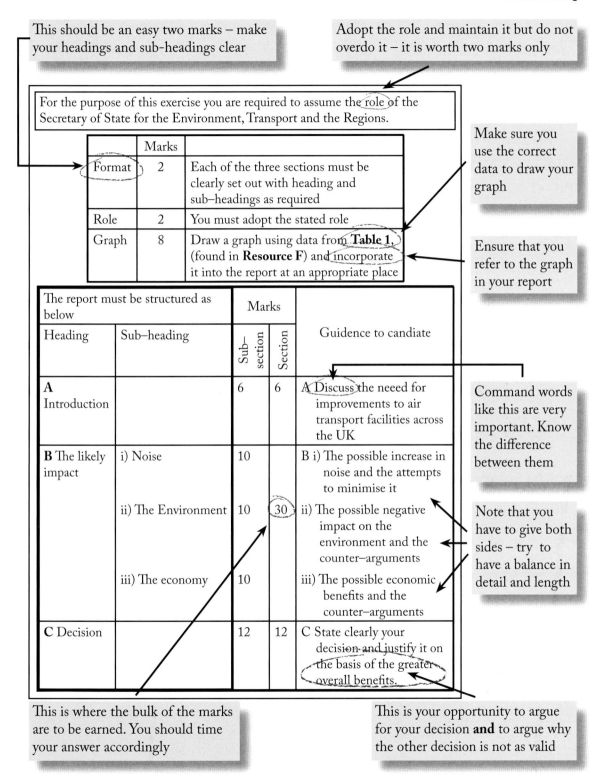

For the purpose of this exercise you are required to assume the role of the Secretary of State for the Environment, Transport and the Regions.

Make sure you use the correct data to draw your graph

	Marks	
Format	2	Each of the three sections must be clearly set out with heading and sub–headings as required
Role	2	You must adopt the stated role
Graph	8	Draw a graph using data from **Table 1**, (found in **Resource F**) and incorporate it into the report at an appropriate place

Ensure that you refer to the graph in your report

The report must be structured as below		Marks		Guidance to candiate
Heading	Sub–heading	Sub–section	Section	
A Introduction		6	6	A Discuss the neeed for improvements to air transport facilities across the UK
B The likely impact	i) Noise	10		B i) The possible increase in noise and the attempts to minimise it
	ii) The Environment	10	30	ii) The possible negative impact on the environment and the counter–arguments
	iii) The economy	10		iii) The possible economic benefits and the counter–arguments
C Decision		12	12	C State clearly your decision and justify it on the basis of the greater overall benefits.

Command words like this are very important. Know the difference between them

Note that you have to give both sides – try to have a balance in detail and length

This is where the bulk of the marks are to be earned. You should time your answer accordingly

This is your opportunity to argue for your decision **and** to argue why the other decision is not as valid

An example of a question from a Decision Making Exercise

5.2 Structure

The **structure** of the Decision Making Exercise will also remain little changed from year to year with an introduction, a section on the impact of the development and a conclusion.

Introduction

You are required to introduce the issue. It is important that you follow the wording of the question here as it is likely that there will be specific aspects of the issue that are required for the introduction. It may be that you will have to use a number of resources to find out about the issue and to address the question. The introduction may ask you to set the proposed development in a context. Sometimes this is a national context – in other words you might have to explain the importance of the development for the *country* in which it is to be built. In that case you must look for those resources that indicate the impact on the country as a whole and use those in your answer. In other cases they may be looking for a regional context and in that case you have to find different resources. The introduction may not be looking at the importance of the issue at all but may, for example, want a description of how it has arisen. The key here, as elsewhere, is to read carefully the requirements of the introduction in the question paper and to spend some time making sure that you are clear about what is required.

Impact of the development

Typically this section is worth the most marks and, if so, you should spend most time on it. It will be divided up for you into subsections. These will vary according to the issue being addressed but are likely to be about the economic, social and environmental consequences of the proposed development. You will probably be required to look at the benefits and the drawbacks of each of these. It is worth reminding ourselves of what we might have to consider in each of these.

> ### Economic
> Will the development bring jobs to the area, increase the earnings and the wealth of the area? Will it cause job losses elsewhere? Will there be a cost in other forms of economic activity?
>
> A quarry, for example, will bring more jobs to an area but might also cause a reduction in jobs in tourism, if it is unsightly or polluting.

Social

Will the proposed development have a positive impact on the people in the area, other than providing them with jobs (an economic benefit)? Will it provide them with more facilities? Will it improve the infrastructure – roads, schools, power supplies, and so on? Will it encourage people to remain in the area and will it improve the quality of life? On the other hand it could have a negative social impact on people if it were to go ahead.

Environmental

Will the development have an impact on the natural environment? This might be pollution of some form (visual, water, air for example) or threats to habitats of native plants and animals. Many proposed developments will claim that they are protecting or even improving the environment or at least minimising the negative impact that they might have.

It is likely that you will be asked to look at the benefits and the drawbacks of the proposed development going ahead. Sometimes you are required to look at the arguments for and the arguments against the proposal. For example, you might have to discuss the pollution which it is claimed the development will cause and then look at the arguments that claim that the pollution threat is unfounded.

Each Decision Making Exercise will be different in its requirements, depending on the actual issue. Some issues may seem to have a limited social impact and so, in that case, it is unlikely that that will form part of the question. Alternatively, in some cases an aspect will have so much material that it might be divided into two sections – for example, the environment may be divided into noise pollution and visual pollution. Look very carefully at the subsections here to ensure that you are answering the question fully.

Do remember that you are given **all** the information that you need to provide each answer – you only have to bring with you 'critical understanding'. The knack is to select efficiently all of the relevant facts that you require in each section (some candidates favour the use of different coloured highlighters to help in this task) and to demonstrate an understanding of the information in your presentation of it. As you cannot just repeat the original text, you often have to reconstruct the information provided in a different form, while retaining the sense of the original. This means that you need to develop an understanding of the issue as rapidly as possible. Practice using past papers is often the best way of doing this.

Conclusion

In this section you are normally required to state your decision and to justify it on the basis of 'the greater overall benefits'. This could be a repeat of the previous section if you are not careful. Do not go over the ground you have covered before. Instead this is an opportunity to compare the arguments and to weigh them up against each other. You could start your justification with "The development company claim that 250 jobs will be created in this development. However local people point out that few of those will be available to them and those that are will be low paid." Alternatively you could argue "While local people fear that few of the jobs created in the development will be available to them, the development company have stated that there will be 250 jobs in all and this will give a considerable boost to the local economy". While these examples lack the detail that you could provide in a real situation, they do show one argument being set against another, which is exactly what is required in this section.

Other aspects of structure

Rôle – you are generally provided with 1 or 2 marks for this and the question will require you to 'adopt and maintain' a rôle. A glance at a Mark Scheme for past papers will show that a candidate will be awarded one mark when they *adopt* the rôle on the first occasion ("As Secretary of State, I …") and a second mark, where available, for the second occasion in a different section when they *maintain* the rôle. Do not write a long covering letter as that wastes valuable time and may earn only 1 mark. Also make clear that it is you playing the rôle, rather than you as a candidate who is writing. "I do not want the development to go ahead because …" will *not* be considered to be adopting the rôle .

Format – you are rewarded if you follow the instructions for this, usually by using clearly the headings and sub-headings provided. Do this by using the wording of the headings and sub-headings as provided and lay these out clearly so that they stand out from the body text. You could underline them, or put them in upper case, and put them on a separate line from the rest of your answer. Also, leave a space between each section.

Graph or other skill – at some stage you are required to demonstrate a geographical skill. This is generally drawing a graph using data provided.

5.3 Techniques

Timing

Good timing is essential in this, as in any other examination. You should look at the allocation of marks within the question and divide your time accordingly. Very often markers of A level papers comment on candidates penalising themselves by omitting sections of the paper through poor timing. In A2-3 this is sometimes the graph being left out because of time constraints. Sometimes too little time is left to do justice to an important part of the question. Often examiners can spot a candidate who is likely to suffer from time problems later in the paper if they have written unnecessarily long answers to questions early in the paper.

Pay attention to the mark allocation within questions. You should have familiarised yourself with the structure of the paper beforehand. You should make a calculation of timing before you start, using the instructions on the front cover of the paper to help you. You can do this in advance for the Section A/Section B time allocation following the recommendations on the front cover of all A2-3 papers. For parts of questions within each section you should make a calculation before you begin each section. Check these calculations very carefully as an error here could be very costly.

WARNING FROM THE CHIEF EXAMINER: NOT COMPLETING ALL PARTS OF YOUR EXAMINATION PAPER WILL COST MARKS!

You are recommended to spend about 25 minutes reading the questions and selecting the appropriate information before you start your answer. This should leave about 70 minutes for the answer itself. The DME is worth 60 marks which equates to about 1 minute per mark, with a little leeway. If, for example, a question is allocated 12 marks, you should spend about 12 minutes answering it. If you can answer it effectively in less time, then that is a bonus. However, it is most unlikely that you could provide an effective answer in 6 minutes. Another mistake would be to spend 16 minutes. That would be likely to put you into time problems later on and, in any case, spending 16 minutes is unlikely to gain more marks.

The graph, if it is required, means that you have to select a suitable graphical technique – a line graph, bar graph or whatever – and use it to represent some data provided in the paper. This is typically allocated 7 or 8 marks. However it is possible that this will take a little longer than 7 or 8 minutes to complete so a small amount of additional time has to be allowed for this – probably some of the additional 10 minutes leeway discussed earlier. It is a serious mistake to omit this or other parts of the task as this will impact on your marks and could cost a grade or more.

Skills

The skills section of the DME has most often required you to show that you can represent in a graph some information provided in a table in the resource booklet. You know a number of graph types including proportional, line, bar and pie.

Only use a line graph when time or some other continuous measurement is along one axis. For data when figures are in separate categories, a bar graph is usually more appropriate. Only in Table a, below, would a line graph be appropriate.

Table a: Revenue from Tourism in Northern Ireland (£million)

1997	268
1998	280
1999	322
2000	329
2001	348

Table b: Distance travelled to shop (kms)

0–10.9	147
11–25.9	23
26–50	6

Table c: Land use in a town (%)

Residential	58
Industrial	10
Retail	21
Derelict	8
Vacant	3

Think carefully before you select the type of graph that you are using to ensure it is the appropriate technique.

Pie charts and proportional graphs are very effective ways of showing information but will probably take too long to construct in the short time available in the examination, so it is likely that your graph will be a bar graph or line graph.

Correct placement of the diagram within the appropriate section is often rewarded as is making reference to the diagram in the body of your answer ("As the graph below shows …") If you feel that it would interrupt the flow of your answer to stop and draw the diagram, then leave a space and return to it later. Do ensure that you leave enough time to do this.

Many marks are lost by not having the usual conventions on a graph. Use something like the following mnemonic and checklist to ensure that you have not left anything out.

T I have given the graph a correct **title** ☐

R I have made a **reference** to the graph in the body of my answer ☐

A I have labelled the **vertical axis** fully (including units eg tonnes) ☐

A I have labelled the **horizontal axis** fully (including units eg years) ☐

K I have drawn a correct **key**, where necessary ☐

Decision

As you worked your way through the materials in the Decision Making Exercise, you may have determined which decision you are going to make ultimately, when you come to the conclusion of your answer. Most people suggest that you should make up your mind very early on what your decision is to be. If you have decided early on in the process, you can be marshalling your arguments for the conclusion as you sift through the material for other sections. You can also retain material which is more appropriate for the conclusion and not have gone into the intricacies of them in the impact section.

Mind mapping

Some people are helped in their visualisation of an issue by drawing a mind map which can help to tease out the aspects of the issue. If you are using this technique in an examination be careful not to make it too detailed as that might pose time difficulties. All examiners have come across candidates who have spent much too long on a plan of their answer, only to fail to complete the answer itself.

The diagram overleaf gives an example of a mind map that might have been drawn by

Mind map for Dibden Bay

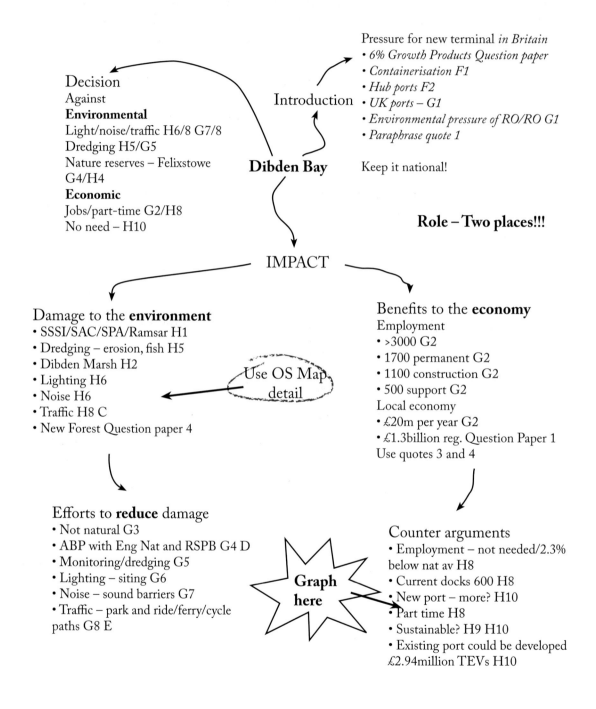

Pressure for new terminal *in Britain*
- *6% Growth Products Question paper*
- *Containerisation F1*
- *Hub ports F2*
- *UK ports – G1*
- *Environmental pressure of RO/RO G1*
- *Paraphrase quote 1*

Introduction

Keep it national!

Dibden Bay

Decision
Against
Environmental
Light/noise/traffic H6/8 G7/8
Dredging H5/G5
Nature reserves – Felixstowe
G4/H4
Economic
Jobs/part-time G2/H8
No need – H10

Role – Two places!!!

IMPACT

Damage to the **environment**
- SSSI/SAC/SPA/Ramsar H1
- Dredging – erosion, fish H5
- Dibden Marsh H2
- Lighting H6
- Noise H6
- Traffic H8 C
- New Forest Question paper 4

Use OS Map detail

Benefits to the **economy**
Employment
- >3000 G2
- 1700 permanent G2
- 1100 construction G2
- 500 support G2
Local economy
- £20m per year G2
- £1.3billion reg. Question Paper 1
Use quotes 3 and 4

Efforts to **reduce** damage
- Not natural G3
- ABP with Eng Nat and RSPB G4 D
- Monitoring/dredging G5
- Lighting – siting G6
- Noise – sound barriers G7
- Traffic – park and ride/ferry/cycle paths G8 E

Graph here

Counter arguments
- Employment – not needed/2.3% below nat av H8
- Current docks 600 H8
- New port – more? H10
- Part time H8
- Sustainable? H9 H10
- Existing port could be developed £2.94million TEVs H10

a candidate who was against the development at Dibden Bay in a recent DME (2003). The structure of the diagram is based on the question and the mind map could be used both to summarise the relevant information and to provide a plan for answering the question. This diagram has been written so that someone else can make some sense of it. In this case the candidate has produced the mind map structure and, while going through the resource booklet, has noted points on the mind map. He or she has numbered the paragraphs in the resource booklet and gives letters and paragraph numbers on the plan to allow easy retrieval of information from the resource booklet when the answer is started.

If you were to use this technique, remember that it is only for use by you. It is not a problem if only you can decipher some of the abbreviations on it for example. If you find this technique too time consuming (and you should practice it using past papers to find out if it helps you), then find another way of visualising the issue, summarising the main points and planning your answer. While the mind map, or other technique, is for you, an examiner may give some credit for the 'plan', if there is some difficulty with time.

Levels of response mark schemes

It is important to know how your questions are being marked so that you can answer in a way that will make your work 'score' as much as possible. All longer questions are marked using Levels. There are usually three of these and the details of them for each question are provided each year in the detailed mark scheme for the paper that your teacher will have. A typical version is shown below.

You do not have to write more to get to Level 3; you just have to show a clearer understanding of the points

Make sure that you do not bring in material that is not relevant to that section

Level 3 (11-15)

Candidate states clearly the main argument and the counterargument. Two or more factors should be discussed. The points made will be consistently relevant and logically structured and show an insight into the issue. Most of the relevant resource material will be used and no significant points will be omitted. Figures where available and appropriate will be used effectively.

If arguing two sides to an issue, try to make sure they balance in length and detail

There is no requirement to use all of the resources, but most will have a role to play in your answer

Level 2 (6-10)

Fewer lines of discussion but, while ideas may lack detail or depth, they are still adequate. There may be a heavy imbalance between the two sides of the argument and there may be an over-reliance on direct quotes from the resources. Some of the resources will not be as well used as they might be and some irrelevant material may be introduced.

Do not copy out the resources – make sure you put the arguments in your own words

Do not use words like 'a lot of' or 'few' if you can use figures from the resources instead

Level 1 (1-5)

Simple understanding is shown but the argument is sketchily dealt with. The resource material is used but many relevant resources are omitted and there is little structure or logic in the ordering of content.

You need to use the detail in the resources to show your understanding of the issue

Making a plan or mind map is essential

When you are able to distinguish between a Level 2 and a Level 3 answer, for example, you are much closer to being able to provide a Level 3 response yourself. You should practice marking parts of your own and others' answers, allocating them to levels and giving them a mark within those levels.

Exam Tip

You might imagine that you cannot prepare in advance for an unseen decision making examination. For example you do not know whether the decision is to be about a harbour development in Germany or a road widening scheme in Scotland.

It is true that you do not know the issue with which you will be faced nor its location. You do not know the full details of the task you have to do or the rôle that you will be asked to adopt.

Nevertheless there is **a lot** you can do in advance of the examination.

- Review the Decision Making Exercises you have done with your teacher, including past papers where available.

- Revise the data presentation skills and the checklist for graph conventions.

- Examine the front page of a recent Decision Making Exercise examination to become familiar with the mark allocations and the time available. You will still have to do this in the Examination Hall but prior thought should reduce the likelihood of a mistake being made.

- Reflect on the levels of response mark scheme – how can you maximise your marks?

- Remind yourself how important good timing is and also how vital it is not to omit any part of the examination paper.